T0310776

TOXIC TOWN

Toxic Town

IBM, Pollution, and Industrial Risks

Peter C. Little

NEW YORK UNIVERSITY PRESS
New York and London

NEW YORK UNIVERSITY PRESS
New York and London
www.nyupress.org

References to Internet websites (URLs) were accurate at the time of writing.
Neither the author nor New York University Press is responsible for URLs that
may have expired or changed since the manuscript was prepared.

Library of Congress Cataloging-in-Publication Data
Little, Peter C.
Toxic town : IBM, pollution, and industrial risks / Peter C. Little.
pages cm
Includes bibliographical references and index.
ISBN 978-0-8147-6069-7 (cloth : alk. paper) — ISBN 978-0-8147-7092-4 (pbk. : alk. paper)
1. Hazardous waste sites—New York (State)—Endicott. 2. Hazardous waste site
remediation—New York (State)—Endicott. 3. Endicott (N.Y.)—Environmental
conditions. 4. International Business Machines Corporation. 5. Computer industry—
Environmental aspects—New York (State)—Endicott. 6. Computer industry—Waste
disposal—Environmental aspects—New York (State)—Endicott. I. Title.
TD181.N72E535   2013
363.738'49—dc23   2013039340

New York University Press books are printed on acid-free paper, and their binding materials
are chosen for strength and durability. We strive to use environmentally responsible
suppliers and materials to the greatest extent possible in publishing our books.

Manufactured in the United States of America
10 9 8 7 6 5 4 3 2 1
Also available as an ebook

For Jenny, Lucy, and Theo

CONTENTS

FIGURES AND TABLES

One simple way to assess the impact of any organization is
to answer the question: how is the world different because
it existed?
—Samuel J. Palmisano, IBM Chairman

On June 16, 2011, International Business Machines Corporation (IBM)
celebrated its one-hundredth birthday and commemorated this event
with a year-long series of seminars and conferences around the world.
The centennial celebration was an opportunity for IBM to reach out
to many of its constituents in industry, government, and academia
and engage with them in a variety of celebratory events. IBM Chairman Sam Palmisano even invited these "members" of the IBM family
to pledge at least eight hours of community service during 2011 to com-
memorate IBM at 100.

IBM has much to celebrate. After a hundred years of business and
technological innovation, the company has a dazzling resume, making investments and impacts in almost every sector of industry, government, and science. They have developed a supercomputer system
named WATSON, named after IBM's founder, Thomas J. Watson, Sr.,
that is able to approximate human speech and respond to questions
with a precise, factual answer by running complex analytics. On a *Jeop-
ardy!* match televised in February 2011, WATSON beat two champions
of the game. IBM even has its own Institute for Electronic Government and is at the forefront of developing electronic voting services
and technologies to advance the worldwide growth of "electronic" or
"digital" democracy.[1] IBM's Computational Biology Center has part-
nered with National Geographic to arrange the largest human genetic

dataset to model and reconstruct the "genographic" spread and makeup of humanity on planet Earth. In 2004, IBM launched Blue Gene, one of the fastest supercomputers in the world, which can compute 91.29 teraflops per second. To put that computing speed into perspective, one teraflop equals one trillion operations per second.

IBM's involvement in the public health and healthcare sectors is also impressive. They have recently worked closely with the National Institutes of Health to develop a massive database of chemical data extracted from millions of patents and scientific literature to allow researchers to more easily visualize important relationships among chemical compounds to aid in drug discovery and support advanced cancer research. They have developed the Spatiotemporal Epidemiological Modeler, which is being used by public health scientists worldwide to trace the spread and distribution of disease. IBM is also at the cutting edge of medical imaging technology, which is improving medical diagnosis, allowing patients to better "see" the internal conditions of their body under medical analysis, and providing families with high-resolution images of their yet-to-be-born child.

IBM is also a major player in the fast-growing field of nanotechnology, which is based on the use of atomic and molecular scale structures and devices for enhancing information technologies. To further augment nanotechnology development and investment opportunities, IBM has generated a new "business" relationship with New York State. In 2008, a research and development mission co-sponsored and financed by IBM and New York State aimed at accelerating New York's international leadership in nanotechnology research and development and creating new high-tech jobs. This corporate-state partnership has led to the development of the Computational Center for Nanotechnology Innovations, a partnership among Rensselaer Polytechnic Institute, IBM, and New York State that is building one of the world's most powerful university-based supercomputers that is further miniaturizing devices in electronics and opening up the capacity of nanotechnology to influence multiple industries. A brand name of American capitalism, IBM even spearheaded the development of barcode technology to track commodities and improve global product distribution.

The list of IBM's contributions to business, science, and technology and its achievements in all sectors of government and industry goes on and on. The company has also received a number of awards for its environmental leadership in the business world. In 2010, it received several "environmental" awards. IBM ranked number one in the *Newsweek* "Green Ranking" of the biggest publicly traded companies in developed and emerging world markets for its business attention to environmental impact, green policies, and reputation. The company ranked number one in the latest Supercomputing Green500 List announced by Green500.org, an organization that ranks the top 500 supercomputers in the world by energy efficiency. The list the organization generates shows that 13 of the top 25 most energy-efficient supercomputers in the world today are built on IBM high-performance computing technology. Finally, IBM ranked number ranked number one in all-around performance and was in the top three in all five categories in Gartner/World Wildlife Fund's recent "Low-Carbon & Environmental Leadership Findings Report," a report that evaluated 28 information and computer technology companies on their all-around performance, including, but not limited to, the company's "internal environmental performance."[2]

Amid these things to celebrate, these products and moments of high-tech prowess and progress, certain consequences of such IBM productions do exist. Lenny Siegel and John Markoff, in their book, *The High Cost of High Tech: The Dark Side of the Chip*, exposed the "toxic time bomb" of the so-called green high-tech industry, writing confidently that "[h]igh-tech pollution is a fact of life wherever the industry has operated for any length of time, from Malaysia to Massachusetts" (Siegel and Markoff 1985:163). The following book extends this effort of exposure by focusing on a case study concerning the material environmental impacts of IBM technology which call into question IBM's new technoscientific ideology based on a corporate desire and mammoth effort to "Build a Smarter Planet" in a decade (2010–2020), which IBM calls the "Decade of Smart." "By a smarter planet," according to IBM's corporate website, "we mean that intelligence is being infused into the systems and processes that enable services to be delivered; physical goods to be developed, manufactured, bought and sold; everything from people and money to oil, water and electrons to move; and billions of people to work and live."[3] A Smarter Planet, as it appears here, is a cosmospheric innovation whereby "everything" can become better with more infusion of "intelligence"—that always slippery term—and the planet *becomes* smarter when IBM's intelligence infuses everything in our socio-natural world.

IBM's Smarter Planet mantra, stated best by Samuel J. Palmisano, IBM chairman and former president, during a speech to business and civic leaders in London on January 12, 2010, is powered by the fact that "Enormous computational power can now be delivered in forms so small, abundant and inexpensive that it is being put into things no one would recognize as computers: cars, appliances, roadways and rail lines, power grids, clothes; across processes and global supply chains; and even in natural systems, such as agriculture and waterways."[4] In that speech he spoke of how digital devices—soon to number in the trillions—are being linked via the Internet, suggesting we are living among an "Internet of Things," resonating with the notion that contemporary reality is one marked by "vibrant materiality" (Bennett 2010). IBM's Smarter Planet initiative is modeled on the mission of using digital technology solutions to modernize the infrastructure of nation-states, by installing energy-efficient (or smart) grid technologies and reducing traffic congestion in cities via sensor technology. It is no secret that IBM is becoming more entangled in city- and nation-building efforts, and is also a senior firm that is representative of and for constantly expanding techno-modernization with deep domestic and international government and industry ties.[5]

IBM—famously nicknamed "Big Blue"—is excited about the capacity of data, what it calls "the knowledge of the world," "the flow of markets," even "the pulse of societies," to be transformed into intelligence, and the capacity of computers to have the processing power and advanced analytics to "make sense of it all."[6] It is with this cutting-edge knowledge and technology that IBM believes it "can reduce cost and waste, improve efficiency and productivity, and raise the quality of everything from our products, to our companies, to our cities."

On the techno-economic surface, this all sounds promising and "smart." But, deliver this message of corporate progress and prowess on a loudspeaker in downtown Endicott, where IBM planted the seeds for its empire of business and innovation, and chances are the message falls on ears of distrust, discomfort, and even disgust.

How might this citizen response, in this community, complicate the story of IBM modernization, the corporate story of ever-increasing processing power and computer analytics? The lived experiences and discourses of Endicott residents explored in this book do not fully answer this question. The book does, on the other hand, argue for a much needed pathway of understanding rooted in ethnographic research and description to show how and why this message of high-tech progress, dynamism, and "smartness" is contradictory and might be contested and thwarted for good reasons. In the spirit of one Endicott resident and former IBMer who has struggled to live a comfortable life in IBM's contaminated birthplace, the following words mark the real active ingredient of this book: "I just want people to know. People need to know that there is a problem here. People need to know that it ain't gonna be covered up. They try to make everything look nice by mitigating, but there is a big problem here. The fact is IBM took a dump on this community. They messed up here, big time."

What follows is one anthropologist's attempt to honor this recognized "problem" and, perhaps more important, to honor the lived uncertainty and frustration that persists after the toxic "problem" has been re-coded as a "mitigated" problem. It documents a community's transformation from being an industrial boom town to a community strangled by the moral hold of "responsible" mitigation. It honors the elusive and precarious nature of risk mitigation in a polluted industrial birthplace. As the French philosopher Michel Serres might put it, it discusses pollution mitigation and repair as one of many extensions of toxic sludge, as a pollution response, reflex, and example of "corporate responsibility" that is as much about risk control and scientific assurance as it is a practice of re-appropriating and re-defining a troubled high-tech birthplace and landscape of late industrial corrosion.

ACKNOWLEDGMENTS

This book would not have come into being without the help and guidance of family, friends, and colleagues. First, the unforgiving love and support of my parents, Nancy and Chuck Little, needs to be acknowledged. They have always encouraged me to do good, work hard, and stick with my passions and morals. This book is my attempt to do just that.

Professors and former graduate students in the Department of Anthropology at Binghamton University, especially Deborah Elliston, Ann Stahl, Carmen Ferradas, Daniel Renfrew, and Thomas Pearson, helped shape my undergraduate honors thesis project on the IBM-Endicott contamination conflict, which planted the seed for the present work. I also wish to thank faculty and friends—you know who you are—in the Department of Anthropology at Northern Arizona University. I thank Miguel Vasquez, Jill Dubisch, Cathy Small, and Bob Trotter. A special thanks to David Seibert for his friendship over the years and for encouraging me to stick with it.

I am indebted to Bryan Tilt, for his mentorship and friendship during my years as a doctoral student in the Anthropology Department at Oregon State University. I thank Nancy Rosenberger, Anna Harding, and David Sonnenfeld who provided insight and guidance on earlier versions of this manuscript. Also, I thank my fellow graduate students and friends at Oregon State University, especially Christina Package,

Courtney Everson, Kai Hennifin, Adel Kubin, Marco Clarke, Daniel Hunter, and Eddie Schmidt. I also thank the following anthropologists for their friendship, support, and continued inspiration: Alexa Dietrich, Barbara Rose Johnston, Melissa Checker, Leah Horowitz, Kim Fortun, Merrill Singer, Britt Dahlberg, Peter Rudiak-Gould, and Valerie Olson.

I thank the National Science Foundation's Decision, Risk, and Management Sciences program for providing the grant (NSF DIG S1096A) necessary to generate this book. I am also grateful for a visiting researcher appointment granted to me by the Department of Anthropology at Binghamton University during my 2008–2009 fieldwork. Thanks also go out to the Anthropology/Sociology Department at the State University of New York at Cortland for hiring me to teach an anthropology course during my field season.

Portions of chapter 6 were published previously by Taylor and Francis as "Environmental Justice Discomfort and Disconnect in IBM's Tainted Birthplace: A Micropolitical Ecology Perspective," *Capitalism Nature Socialism* 23(2):92–109. Portions of chapter 7 were published previously by the Society for Applied Anthropology as "Vapor Intrusion: The Political Ecology of an Emerging Environmental Health Concern," *Human Organization* 72(2):121–31. Portions of chapter 5 were published previously by the American Anthropological Association as "Another Angle on Pollution Experience: Toward an Anthropology of the Emotional Ecology of Risk Mitigation," *Ethos* 40(4):431–52. I thank these presses for their permission to republish this material.

I thank Christina Geller and Conor Maguire for producing the study location map, as well as Doug Jongeward for his technical assistance with images. Special thanks also go to the IBM Corporate Archive for their permission to use selected images from their image archive. I wish to thank my editor at NYU Press, Jennifer Hammer, for her friendly spirit, clarity, and generosity in guiding this book to completion.

Heartfelt thanks go to Jenny Little—my partner every step of the way—for her patience and support during the research and writing of this book. Without her and our children, Lucy and Theo, by my side each day, I would surely be caught in a vortex of madness and confusion. Their smiles, laughter, and love inspire me to enjoy life and I am glad this book has left the home and ventured into the public sphere, if not simply to allow us to have a few more hours each day together.

Last, I wish to thank all the people of Endicott and beyond who participated in this study. I am especially indebted to the residents, activists, scientists, regulators, and public officials who shared and opened their minds and homes to me throughout the course of this project. To maintain confidentiality, I can't thank them by name, but they know who they are. I am grateful for their willingness to share their perspectives, experiences, knowledge, and stories, and I apologize in advance if this book misrepresents your story of struggle. I have used pseudonyms throughout the book to protect informants' identities as best I could, and I take full responsibility for any misinterpretation found throughout.

ATMS	Anti-Toxics Movements
ATSDR	Agency for Toxic Substances and Disease Registry
CARE	Citizens Acting to Restore Endicott's Environment
CCIA	Computer and Communications Industry Association
CERCLA	Comprehensive Environmental Response, Compensation, and Liability Act, more commonly known as Superfund
CVOC	chlorinated volatile organic compound
IBM	International Business Machines Corporation
IRM	Interim Remedial Measures
ITRC	Interstate Technology and Regulatory Council
NIOSH	National Institute for Occupational Safety and Health
NYSDEC	New York State Department of Environmental Conservation
NYSDOH	New York State Department of Health
EJ	Endicott-Johnson Shoe Company
EPA	U.S. Environmental Protection Agency
OSHA	Occupational Safety and Health Administration
PEM	Political Ecology of Mitigation
RAGE	Residents Action Group of Endicott
RCRA	Resource Conservation and Recovery Act
ROD	Record of Decision
TCE	trichloroethylene
VI	vapor intrusion
VOC	volatile organic compound
VMS	vapor mitigation system (or sub-slab depressurization system)
WBESC	Western Broome Environmental Stakeholders Coalition

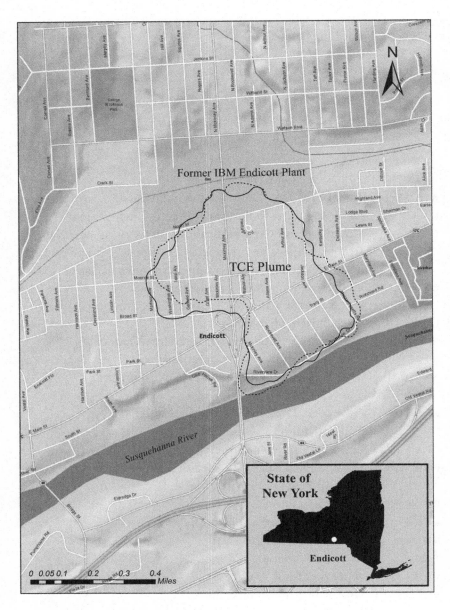

Figure 1.1 Map of Study Area

1

Down in Big Blue's Toxic Plume in Upstate New York

I say, thank God IBM is here to take care of this mess.
—Former Mayor of Endicott

Don't mitigate me and tell me everything is ok.
—Endicott resident

In September 2008, I was doing fieldwork in Endicott, New York, the site of both IBM's first manufacturing plant and a contentious U.S. Environmental Protection Agency Superfund[1] site consisting of a 300-acre toxic plume of trichloroethylene (or TCE), which is a cancer-causing chlorine-based cleaning solvent heavily used by IBM to manufacture chipboards and other microelectronics. I was sitting down with Tonya, a resident of what is locally referred to as "the plume," when I first sensed the need to take the concept of mitigation more seriously. Tonya was sitting on the edge of her couch and seemed excited to talk, explaining from the start of the interview that her two boys had moved out two years ago, and that it was nice to have a visitor. Before I asked my first question, she interrupted with, "You mentioned that you wanted to talk about the IBM spill, well just listen." She pointed across the living room to the western wall of the house. "That's it," she says. "I can hear it going all the time. You forget about it, but you also always know it is there."

Tonya was referring to a ventilation system that runs along her chimney stack on the outside wall of the house. It is also known as a vapor mitigation system. Why this humming mitigation system is on Tonya's house and what it might mean to live in a "mitigated" home in a contaminated neighborhood in a deindustrialized town is a question that, as this book illuminates, opens up a can of worms, for it calls for a perspective on socio-environmental experience that discerns the tangle of social, political, economic, and scientific forces shaping situations and events of technological disaster.

Endicott is in what is known as the "Southern Tier" region of upstate New York and is about three hours northwest of the New York City metropolis. Incorporated in 1906, Endicott is a village in the town of Union and is located in western Broome County along the Susquehanna River. According to the U.S. Census Bureau 2010 Census, Endicott had a population of 13,392, with the majority of residents (86.6%) identifying themselves as white.[2] Industry enthusiasm has long been an active pulse of the community, and most especially, at one time at least, Endicott was a proud supporter of International Business Machines Corporation (IBM), which opened its first plant there in 1924. As the "birthplace of IBM," residents know the local IBM plant played a significant role in the development of the Computer Age. In a Reagan-Bush rally speech on September 12, 1984, Ronald Reagan stood at a podium at Endicott's high school football field holding a jersey imprinted with his nickname, "The Gipper," reminding the people of Endicott about their "valley of opportunity," their valley of prosperity amid an emergent era of American economic policy marked by, among other things, rampant deindustrialization (Bluestone and Harrison 1982). In Reagan's words:

> This valley became home to some of the proudest communities in our nation, towns that had seen firsthand all that free men and women can accomplish . . . [O]ne group of men and women had a great vision, a vision to bring this valley prosperity it had never before dreamed possible, a vision to launch a revolution that would change the world. . . . Their leader was Thomas Watson, Sr. He had grown up in a small town called Painted Post, down the road from here, where he learned how to stick with a job until it's finished. . . . In 1953 . . . the company that Watson had renamed IBM began making the first mass-produced commercial

computer in history—the IBM 650—less than half a mile from this spot. . . . When IBM began, the best market researcher predicted that fewer than 1,000 computers would be sold in the entire 20th century. Well, IBM's first model sold almost twice that number in just 5 years, and now there are IBM plants in Endicott and around the world. And the computer revolution that so many of you helped to start promises to change life on Earth more profoundly than the Industrial Revolution of a century ago. . . . Already, computers have made possible dazzling medical breakthroughs that will enable us all to live longer, healthier, and fuller lives. Computers are helping to make our basic industries, like steel and autos, more efficient and better able to compete in the world market. And computers manufactured at IBM . . . guide our space shuttles on their historic missions. You are the people who are making America a rocket of hope, shooting to the stars. . . . Today, firms in this valley make not only computers but flight simulators, aircraft parts, and a host of other sophisticated products. (Reagan 1984)

Eventually, the high-tech industry's culture of obsolescence (Slade 2006) caught up with the IBM Endicott plant, resulting in "sophisticated" downsizing and deindustrialization, a topic closely linked to neoliberal political and economic restructuring starting in the late 1970s (Harvey 2007, 2005; Zukin 1991). In 2002, 18 years after Reagan's speech, IBM's Endicott plant, which at its peak employed 12,000 workers, was for sale, and residents began to receive invitations in the mail to attend "Public Information Sessions" organized by IBM and state agencies to learn about IBM's efforts to clean up a groundwater contamination plume (or pollution zone) it was leaving behind. Many of these early information sessions were held at the Union Presbyterian Church in west Endicott, just four blocks from where I lived at the time. I found myself amid this IBM contamination debate as both a resident and a budding anthropologist completing my undergraduate degree at Binghamton University, just six miles to the east. I was as much confused and concerned as a resident as I was interested and curious about the intersections of high-tech production, pollution, and environmental public health politics.

All Endicott residents were invited to these "information sessions." They were intended to showcase the collaborative effort of IBM and government agencies (e.g., the New York State Department of Environmental Conservation, the New York State Department of Health, the U.S. Agency for Toxic Substances and Disease Registry, and later, the Center for Disease Control and Prevention's National Institute for Occupational Safety and Health) involved in the oversight of the groundwater remediation effort and the emerging problem of toxics vapor intrusion (the process by which volatile organic compounds migrate from groundwater into overlying buildings).

The first time I attended one of these information sessions, I witnessed two women arguing with a representative of the state health department. Keeping to myself, I moved on to the next information table that had a detailed map of the plume area. After spending some time meeting different agency representatives, I grabbed some cookies and juice, secured the numerous "fact sheets" under my arm, and headed home. As I walked out of the building, I ran into the two women who had been arguing with the health official. I introduced myself and explained that I was a resident and student in anthropology at Binghamton University and was interested in learning more about residents' perspectives on and experiences with the IBM spill. We exchanged numbers, and that interaction planted the anthropological seed in me. It was the starting point of my ethnographic journey, my entry into what forms the focus of this book, which is ultimately how Endicott residents understand and talk about the multiple consequences of the IBM spill; how residents blend contamination and deindustrialization concerns when discussing contemporary Endicott; how residents living in "mitigated" homes above the IBM toxic plume understand and talk about the health risks of toxic exposures and the efficacy of risk mitigation; and what prompted residents to take action and what politics and values did or didn't inform citizen action.

Guided by these questions, this book explores peoples' understandings of, negotiations with, and reactions to high-tech industrial pollution in Endicott, New York, the "birthplace" of IBM. It grapples with the various "dynamics of disaster" (Dowty and Allen 2011) provoked by chipboard manufacturing. This text explores how the IBM contamination conflict in Endicott has been shaped by scientific, political, social, and economic factors locally and beyond. In many ways, the struggle of residents living in the IBM-Endicott plume is a response to transformations in the political economy and ecology of the late twentieth century and the early twenty-first century. As a "Rust Belt" community coping with the pains of industrial pollution, deindustrialization, and other rampant neoliberal consequences of an economy based on the ceaseless concentration of finance capital and protection of shareholder interests,[3] Endicott has become a high-tech "bust town" (Bluestone and Harrison 1982) struggling with various concerns and uncertainties symbolic of our late industrial times (Fortun 2012). These include layoffs, environmental health risk, property devaluation, a dissipated tax base, stigma, corroding factories, toxic solvents, a general decline in the quality of social and economic life, and the expansion of the "discursive power" of corporate social and ecological responsibility (Rajak 2011). At the same time, among environmental scientists familiar with vapor intrusion, the IBM-Endicott contamination site has become an example-setting site for the large-scale mitigation of TCE vapor intrusion (VI). VI is an emerging environmental public health policy concern at state and federal levels as the connections between groundwater contamination and indoor air impact develop.

This book also presents the first study to examine vapor intrusion risk and mitigation from an ethnographic perspective, and to experiment with a mixture of anthropology, political ecology, and science and technology studies (STS) theory to critically diagnose the social, environmental, and health consequences of high-tech production in the American Rust Belt, a region generally associated with the deindustrialization and decay of the auto and steel industries. I take Latour's insight seriously: "[t]o 'study' never means offering a disinterested gaze and then being led to action according to the principles discovered by the results of the research. Rather, each discipline is at once *extending* the range of entities at work in the world and actively participating in *transforming* some of them into faithful and stable intermediaries" (Latour 2005:257).[4] The present study confronts the increasing synthesis of political ecology, science and technology studies, environmental justice, and social theories of risk, to augment and transform the anthropology of technological disaster.

A number of anthropologists have engaged the intersection of capitalist production, pollution, and risk (see Balshem 1993; Button 2010; Douglas 1992; Douglas and Wildavsky 1982; Fortun 2001; Fox 1991;

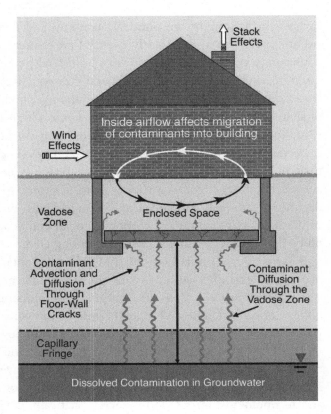

Figure 1.2 Vapor
Intrusion Model.
Courtesy ITRC.

Petryna 2002; Tilt 2004, 2006). Different theoretical frameworks have
turned these researchers on to different research questions and foci, yet
the contentious relationship between industrial toxics and the over-
whelming visible and imperceptible consequences and uncertainties of
contamination continue to draw the attention of social scientists. Risk
and uncertainty invoked by toxic contamination is a sphere of research
that has challenged social scientists to rethink and blend theories to
better understand the social, political, and economic processes stimu-
lating contamination debates and conditioning residents' struggles to
understand often complex environmental exposures (Scammell et al.
2009). In many ways, this book continues this struggle to understand
and uses ethnographic description to contextualize corporate pollution
and community response in IBM's birthplace. Where this book offers

something new is the attention it gives to the topic of "mitigation" and local responses to and negotiations with corporate and state efforts "to fix," to soften the blow or consequences of high-tech industrial pollution. The book critically assesses the efficacy of and social response to toxics mitigation by anchoring the debate in ethnographic description. I don't claim to be telling the "whole" story, nor do I find that even to be the purpose of anthropology or political ecology. Instead the book adds a new circuit, another angle to the telling of Endicott's story of microelectronic calamity. Furthermore, it exposes the paradox of IBM as an "environmental" leader in the business world.[5] For example, as part of its "corporate responsibility" plan, IBM set as a goal of its "environmental affairs" policy the following: "Be an environmentally responsible neighbor in the communities where we operate, and act promptly and responsibly to correct incidents or conditions that endanger health, safety or the environment. Report them to authorities promptly and inform affected parties as appropriate."[6] This book is a report of another kind. Among other things, it addresses how "responsibility" is received, experienced, made sense of by "affected parties," and contested by this affected public. It reports instead on how people's attachments to home, neighborhood, and community have been transformed by irresponsible IBM contamination and deindustrialization, as well as how feelings about dwelling, space, and place are affected by the ecology of senses and understandings provoked by exposure to risks and uncertainties of technological disaster, one of the many persistent themes of late industrialism.

While many of my interests in and thoughts on the IBM contamination conflict have changed since I first confronted the issue as a resident in 2002, much has stayed the same. First, power relations and epistemic turf wars persist, as is common in "contaminated communities" (Edelstein 2004) where corporate, state, and federal "experts" are set in opposition to a local or "lay" public (Gottlieb 2005; Ottinger 2013; Tarr 1996). In other words, my interests in the contested nature of knowledge and claims to knowing the many dimensions of the contamination and its social, economic, and health consequences endure because this struggle has a particular staying power that will likely inform the future of the debates, regardless of shifting decision-making processes. Even where scientists and citizens have sat down at the table together, there is a persistent "uneasy alchemy" (Allen 2003; Ottinger 2013) that surfaces at public meetings, in news reports, and in the ethnographic narratives explored in this book.

I hold another position that has not waned over time. While grass-roots action in Endicott has been "successful" with regard to effectively prodding IBM and government environmental and public health agencies to listen to their concerns and take action, I feel, like many other residents and activists, that no matter what is done to fix the damaged environment and economy of the community, Endicott will never be the place it was when IBM thrived in this region of New York. It is a community of irreversibility in this sense, a chronically de-territorialized and deindustrialized place. There is a clear difference, I was repeatedly told, between Endicott *with* IBM and Endicott *without* IBM. Many residents feel they have experienced IBM abandonment, or what might also be called high-tech capital flight (Cowie 1999), as the tax base of Endicott vanished as soon as IBM began to heavily downsize in the late 1980s and throughout the 1990s. In this way, I will have succeeded in telling Endicott's story of toxic struggle only if the reader comes away understanding and still pondering: (1) how and why local idioms of struggle and frustration *interconnect* deindustrialization, contamination, and risk mitigation experiences; and (2) how and why IBM's social and environmental responsibility track record is open to social critique.

Ethnographic Fieldwork Overview

This book is based on ethnographic fieldwork that took place during 2002–2003 and 2008–2009. During the 2002–2003 fieldwork, I was a resident of Endicott while completing my undergraduate degree in anthropology at Binghamton University. My honors thesis project (Little 2003) was based on semi-structured ethnographic interviews (N=13) with residents I met at the round of Public Information Sessions that began in 2002, soon after IBM sold its Endicott facility. After securing a grant from the National Science Foundation's Decision, Risk, and Management Sciences Program, I returned to Endicott in the summer of 2008 to conduct a more in-depth, ethnographic study to better understand the experience of affected residents. What was different this time around was that IBM, under the pressure of the NYSDEC, had installed nearly 500 vapor mitigation systems (VMSs) on homes and businesses in the 300-plus acre plume. This new landscape, what I describe in detail later as the *mitigation landscape*, added a different twist to the research I had previously done.

In the earlier fieldwork, I was most interested in residents' understandings of the health-contamination relationship as it related to local drinking water quality and risk. But, with the mass installation of these mitigation systems, I decided it was appropriate to explore local environmental health risk perception vis-à-vis the mitigation effort, to explore community understandings of vapor intrusion and the degree to which residents living in the "mitigated" plume area trust IBM and the responding government agencies. What follows is a description of the field methods used to investigate these research interests during the 2008–2009 fieldwork.

I returned to Endicott in June 2008, and my goal was to use a combination of qualitative and quantitative ethnographic methods to help tell the story of local struggle and negotiation in the mitigation phases of the IBM spill (Little 2010). I conducted in-depth, semi-structured interviews with local residents (n=53), including residents living within the defined plume area that encompasses more than 300 acres in the central downtown area of Endicott, and those living outside the plume boundary. These interviews explored how the IBM contamination issue and the mitigation effort are subjectively understood by Endicott residents, local activists, public officials, and regulators and scientists involved in vapor intrusion research in Endicott and nationwide. All interviews were tape recorded with permission and an informed consent document was signed before all interviews. I also did many follow-up interviews by phone and corresponded with activists via email communication. Stratified snowball sampling was used to recruit research participants for this phase of the project (Bernard 2006). My sample was "stratified" or targeted in the sense that I defined "plume" residents as those that lived in the plume, and would get contacts from plume residents about other plume residents. For the activists I interviewed, I targeted residents and community advocates who had direct connections with advocacy groups (e.g., Western Broome Environmental Stakeholders Coalition and the New York State Vapor Intrusion Alliance, and Alliance@IBM) working on the IBM-Endicott issues and who could connect me with others engaged in the network of activists working on these issues.

Another goal of my ethnographic research was to develop a quantitative survey questionnaire that could provide another pathway of characterizing (or visualizing) the experience and perspective of "mitigated" residents living in the plume with vapor mitigation systems. I developed the plume survey instrument in collaboration with members of the Western Broome Environmental Stakeholders Coalition (WBESC). After developing the survey instrument and pre-testing it for clarity, flow, reliability, and validity, I distributed the survey packet (e.g., cover letter, informed consent document, survey, and a self-addressed and stamped envelope) to all homes and residential buildings (e.g., duplexes and apartment buildings) in the plume area that had VMSs (N=464). The purpose of the questionnaire was to record (or trace) residents' demographic information, as well as information about control variables (renter vs. owner, gender, length of residence, education, age, etc.). The questionnaire included scaled questions about residents' sense of environmental health risk, their level of trust in government agencies' response to the IBM contamination, and their level of knowledge of and trust in local risk mitigation efforts. I developed response "values" for these scaled questions, with 1 = strongly agree, 2= somewhat agree, 3= neither agree nor disagree, 4 = somewhat disagree, and 5 = strongly disagree.

As a tool to measure (or visualize) local understandings of the efficacy of risk mitigation technologies and vapor intrusion policy, the survey questionnaire was used primarily to assess residents' understandings of TCE risk and their level of trust in these vapor mitigations systems and the degree to which the installation of these systems on people's homes determines their sense of environmental health risk. Out of the 464 surveys distributed,[7] I received a total of 82 completed surveys, which is a response rate of about 18 percent. This was more than I had originally expected—I was warned by members of the WBESC that since so many plume residents were listed plaintiffs in the Class action, I would have a hard time getting them to fill out anything. As one resident and activist put it, "You see, once the lawyers got involved, there was a lot of hushing up, because the lawyers tell everyone not to talk to anybody. If they win the case, maybe it's a good idea, but it does not look like the case will go that way. IBM will keep in the court as long as they want." In fact, out of 82 respondents to complete the survey, only 24 (or 29.3%) were listed plaintiffs. Also, only 29 (or 36.3%) were former IBMers and only 14 (or 17%) accepted IBM's $10,000 offer to offset the property devaluation caused by the plume, a number that IBM determined was about 8 percent of the home's market value.[8] The majority of plume survey respondents were homeowners (61%) and 39 percent were renters. Females made up 63 percent of respondents and males made up 37 percent. The overwhelming majority identified themselves as white (96%), and 4 percent identified themselves as African American.

I pursued this area of research because community responses to mitigation decisions was an obvious area of research that was being neglected, despite New York State's well-developed and sustained effort to focus on the science of TCE remediation and vapor intrusion in Endicott. Two hypotheses the survey aimed to test were that (1) residents living in homes with vapor mitigation systems have a low sense of risk, and (2) residents with a high sense of risk blame the IBM contamination for personal health problems. I explore these hypotheses, these knowledge gaps, in chapters 4 and 5.

The fieldwork also involved ten interviews with local activists and members of advocacy groups in other communities affected by TCE vapor intrusion. I conducted these interviews to generate qualitative, in-depth ethnographic information on the lived experience of advocates in Endicott and beyond who have become community stakeholders in response to TCE contamination and vapor intrusion. The interviews aimed to expose how, why, and to what extent life experience(s) influence local grassroots action and advocacy. These interview questions included, but were not limited to: Why did you get involved in advocacy work? What would you like to see happen or change to better reflect your own interests and the interests of the advocacy group of which you are a part? What has it been like being an activist? How has your involvement in advocacy changed your life?

In addition to these general questions, I also asked activists in Endicott to reflect on their experience with and perspective on the National Environmental Justice For All tour that visited Endicott in 2006. Of particular interest in these interviews was the question, Do you think of the IBM-Endicott pollution conflict in terms of environmental justice?

As is tradition in the ethnographic methods of anthropology, I used participant-observation research techniques throughout all phases of the project as a way of triangulating qualitative data and increasing validity (Bernard 2006). I lived in Endicott as a resident and researcher (2001–2003, and 2008–2009). Especially in 2008–2009, I had daily, intimate interactions with plume residents. I met residents at public meetings and when I went door-to-door to distribute the quantitative survey. I also met residents who attended the monthly meetings held by the WBESC. I attended all other public meetings regarding the IBM contamination issue, especially those held by the New York State Departments of Environmental Conservation and Health. I took intensive field notes at these public meetings and observed interactions between presenters and meeting attendees. Additionally, I recorded and analyzed archival data on Endicott's social and industrial history at the Endicott Visitors Center, where I spent many days during my 11 months of fieldwork. Endicott's Visitors Center houses all the IBM and Endicott-Johnson Shoe Company archives and has a permanent exhibit documenting the history of Endicott as a "two company town" (see chapter 3).

While in the field and mostly after returning in June 2009, I used *theme analysis* and *content analysis* to analyze and code the narrative data generated from the in-depth interviews (Emerson et al. 1995; Hammersley and Atkinson 1995; Sanjek et al. 1990). These methods were used to strengthen ethnographic interpretations and better contextualize the findings of both the in-depth interviews and the quantitative survey questionnaire.[9]

Anthropologists have traditionally utilized pseudonyms in their ethnographic writing to protect their subjects, and I do the same in this book. That said, all the names that I have used for informants discussed here are the result of my own creation, though this method of "protection" is not perfect. In chapter 6, I use pseudonyms for several activists who have already gone public about their views, but I made every effort to stick with the ethical research goal of protecting the identity of my informants. In this book, I have stuck with this convention, despite my interest in and nod to critical ethnographic reflexivity.

The Chapters in Brief

The book is organized as follows. Chapter 2 lays out my theoretical interests as they relate to the tangle of politics of science, risk, and neoliberalism[10] in communities tainted by technological disaster. I use this chapter to introduce and develop the concept of *mitigation landscape*, discussing how this concept figures in theoretical discussions at the intersection of political ecology and science and technology studies (STS). Next, chapter 3 explores Endicott's industrial history, including a discussion of the development of IBM and the influential role of the Endicott-Johnson Shoe Company. Included in this industrial genealogy is an exploration of IBM's deindustrialization process as it unfolded in Endicott and a discussion of the emergence of the IBM contamination conflict subsequent to the sale of the facility in 2002, which was quickly followed by local grassroots action.

Chapters 4 and 5 draw on ethnographic findings to explore residents' experiences with and perspectives on risk, ambiguity, deindustrialization, community change, and local experiences with and critiques of mitigation. The questions guiding these chapters are: what is the experience of residents living with/in the IBM plume? Amid IBM's boom and bust, and after more than 30 years of remedial work, what do Endicott's plume residents feel certain about? What remains elusive? Guided by these questions, I consider how intersubjective experience informs plume residents' understandings of TCE risk, and draw attention to the dynamics and contradictions of plume residents' understandings of vapor intrusion risk to highlight local environmental health politics. Additionally, because I found that residents' sense of risk was commonly shaped by concerns about occupational health exposures to TCE—particularly concerns about former IBMers who worked at the Endicott plant, which is the focus of an ongoing National Institute for Occupational Safety and Health (NIOSH) study—the chapter also discusses the logical coupling of environmental and occupational health risk, as these spheres of risk are not easily partitioned within the context of toxic plume living. Another goal of chapter 4 is to analyze narrative and survey data elucidating the intersecting themes of IBM deindustrialization, community corrosion, and stigma.

In chapter 5, I draw on narratives from in-depth interviews and quantitative survey data to explore local understandings of the mitigation effort and the lived experience of residents living in "mitigated" homes in the IBM-Endicott plume. After providing background on the state of the art of vapor intrusion mitigation and the ways in which "pro-mitigation" discourse plays out among regulators and activists alike, I analyze plume residents' understandings of vapor intrusion mitigation and the experience of living in a mitigated home by drawing on my interview and survey findings. I discuss, in particular, the theme of persistent or durable ambiguity that one might think would have been softened by IBM and New York State Department of Environmental Conservation (NYSDEC) mitigation efforts. A second goal of the chapter is to engage a topic that has become a popular source of concern for communities threatened by vapor intrusion risk: property devaluation. I analyze this issue as an example of what I call "mitigation nuisance" or a remedial decision or toxics repair tactic that introduces new forms of distress and annoyance for homeowners. But, property devaluation is less a concern for renters, who are a growing population in Endicott's plume. This chapter draws on ethnographic narratives to discuss the effectiveness, or lack thereof, of a tenant notification bill designed to both inform renters about IBM's TCE plume and provide them with results from indoor air sampling of their rental property if those data are available. Renter notification debates, I argue, further intensify the "political ecology of mitigation" (Little 2013a) perspective explored throughout the book.

Chapter 6 draws on activists' narratives and lived experience to better understand the subjectivity or point of view of active community stakeholders in Endicott. With a general focus on the intersections of subjectivity and intentionality, this chapter aims to "make sense" of these citizens' advocacy efforts, and draws on narratives generated from in-depth ethnographic interviews with local activists. Several investigative journalists (Gramza 2009; Grossman 2006) have helped popularize and tell the story of environmental activism in Endicott. This chapter aims to deepen that understanding by drawing on extensive narratives from in-depth ethnographic interviews to highlight activists' perspectives and to illustrate the ways in which this local activism converges with "boundary movement" (McCormick et al. 2003) discourse and action. The chapter also looks at local activism in relation to and in tension with the environmental justice frame and movement. In 2006, the National Environmental Justice for All Tour made a visit to Endicott to connect with local activists, despite the fact that local activism in Endicott was and remains devoid of any environmental justice discourse or intentionality. Some activists I interviewed understood and welcomed the environmental justice "connection" in Endicott, while others found it irrelevant or a stretch. This tour inspired me to explore a critical question: how do we account for environmental justice in communities where it seems discursively vacant and even contested as a legitimate frame? This question challenged me to engage what I am calling "conflicted environmental justice," or what amounts to the volatile and pliant identity politics and "intra-community tensions" (Horowitz 2008) that challenge environmental justice advocacy efforts to pluralize and build a movement "for all."

The focus of chapter 7 is the role of citizens and experts in emerging vapor intrusion debates. The chapter showcases the experience of scientists, community leaders, and regulators engaged in emerging vapor intrusion science and policy, with the goal of illustrating how the toxic struggle in the IBM-Endicott plume goes beyond Endicott and exposes the broader complexities of vapor intrusion science and policy. Many communities across the United States are coping with the threats of vapor intrusion, and Endicott's story is one of many. I use this chapter to share my own prescriptions for democratizing the science and policy of vapor intrusion, showing how as a new and emerging environmental health risk, it is *open* to fruitful anthropological insight and critique.

The book concludes with chapter 8, where I critically reflect on IBM's recent business goal of "Building a Smarter Planet." The IBM-Endicott plume case study, I contend, sets in bold relief the paradox of this ambitious planetary intelligence mission. Such a "global" corporate responsibility effort, one aimed at extending beyond the boundaries of the city and the nation-state, misses critical ecologies of concern and care that matter in the world IBM inhabits and wishes to transform and make "smarter." If what is unfolding in Endicott is an example of a "Smarter Planet," we occupy a planet shaped by an "ethos of technocapitalism" (Suarez-Villa 2009:3) harboring vibrant contradictions and uncertainties, a planet of many difficult landscapes and troubled spaces, even disaster landscapes of perturbed and precarious mitigation.

2

The New Mitigation Landscape

mit-i-ga-tion, *noun*. 1. the act of mitigating, or lessening the force or intensity of something unpleasant, as wrath, pain, grief, or extreme circumstances; 2. the act of making a condition or consequence less severe; 3. the process of becoming milder, gentler, or less severe.

Buildings as landscapes, not persons, were in need of care.
—Murphy (2006:144)

If one looks closely enough, Endicott, New York, best known as the "birthplace of IBM," is today a landscape of toxics mitigation technology. It has become a Computer Age ruin and a place of pollution repair. Venting systems with white plastic tubing running from basements to roofs are visible on nearly 500 houses and businesses in Endicott's downtown area, and were all paid for by IBM. Many residents feel the area has become a taboo space and see it staying that way as long as the toxic plume lurks. The "plume"—a roughly 300-acre zone polluted by industrial toxic substances once used for circuit board production at the former IBM plant—is what I call a *mitigated landscape*. The idea and force behind my use of the concept of *mitigation landscape* comes down to the simple fact that the plume landscape, the IBM-Endicott Superfund site, is one marked by, among other things, mitigation technology that is immediately observable and above the ground, unlike the elusive toxic plume lurking beneath. The contamination of groundwater and soil in Endicott has resulted in the presence of vapors that waft into hundreds of homes. The mitigation of these intruding toxic vapors has turned the plume zone into a zone or landscape of mitigation, a place marked by toxics intervention. Occurring over decades—starting at least in the late 1970s—the contamination, which consists mostly of volatile organic compounds (VOCs), is traceable to IBM chipboard manufacturing. The venting systems, also called vapor mitigation systems (VMSs) or sub-slab depressurization systems, have become a driving symbol of this IBM disaster. They have made risk "public" and viewable, a matter of visible concern, which supports the perceptive that "Without techniques of visualization, without symbolic forms . . . risks are nothing at all" (Beck 2006:332).

In addition to having concerns about the prolonged effects of toxic vapors on the health of the community, local homeowners feel anchored to residential property with no value, they are concerned about

Figure 2.1 Vapor Mitigation System. Photo by Peter C. Little.

devaluation of their investments, and feel they are left to cope with the stigma that goes with living in "the plume." The main contaminant of concern is trichloroethylene (TCE), a known cancer-causing toxic substance that has recently been determined by thousands of epidemiological studies to be "carcinogenic to humans by all routes of exposure and poses a potential human health hazard for noncancer toxicity to the central nervous system, kidney, liver, immune system, male reproductive system, and the developing embryo/fetus" (Chiu et al. 2013:303). Beginning in 2002, the New York State Department of Environmental Conservation (NYSDEC) began investigating TCE vapors at the IBM-Endicott site. Since then, the TCE has been measured at very low levels in many areas of the plume, but this finding does not seem to comfort those worried about several decades of exposure. The ambiguity regarding long-term exposure has been a source of concern for many residents, despite the message from officials and experts that there is a low public health threat. In other words, "risk analysts can calculate till they are blue in the face" (Douglas and Wildavsky 1982:89), but plume residents feel uneasy about living in a place contaminated by TCE, even if their home and neighborhood have been designated a site of successful mitigation. Ambiguity and frustration endure amid mitigation efforts, leaving residents to dwell in an elusive environment of risk and confusion.

The risk mitigation technologies marking the Endicott landscape are, like objects themselves, "many things at once" (Murphy 2006:10) and in relation to many other things at once, such as the social and economic stressors and disturbances of deindustrialization. As one activist in Endicott told a newspaper reporter, "There are two IBMs. . . . In their early days, their philosophy was to take the best care of their employees. . . . By the 80s a new IBM came forward. That's the one that cares about the bottom line." Many residents I spoke with mentioned this troublesome corporate sea change, contending that in the 1960s and 1970s IBM was a responsible company. Many residents I spoke with became almost nostalgic when they recalled the concerts the IBM band would perform for the community, especially during holidays. As one activist resident put it, "We went from being a great place to raise a family to where we are today. Now it [Endicott] has a dark side to it."

The chapters that follow aim to unpack and diagnose this "dark side" or dystopic narrative in a way that sheds light on local residents' experiences of deindustrialization, contamination, and risk mitigation. This thematic tension informs my own struggle to articulate, theorize, and compose this complex high-tech contamination debate as an anthropologist who views theorizing to be a personal experience or experience of personal discovery (Hastrup 1995; Swedberg 2012).[1] I am a researcher of socio-environmental life who thinks about anthropological theorizing as an ethnographically grounded (Nader 2011) practice of articulation revealing "less a set of ideas that somehow mirror the world, and more an assemblage or contraption of different parts, humanly designed within the world to do certain kinds of work" (Fortun and Bernstein 1998:40). Fieldwork not only shapes the researcher, it informs the imaginable.

Political Ecology, STS, and the Mitigation Landscape

While the message expressed by one Endicott resident (see figure 2.2) basically summarizes my own understanding of the contentious issues presented in this book, a certain overarching theoretical move or "thought experiment" (Stengers 2010:12) informs the

Figure 2.2 TCE Is Toxic. Photo by Peter C. Little.

political-ecological analytics deployed here. The perspective of political ecology, with its general deep curiosity with the intersections of global political economy, environmental change, and nature-society dialectics and conflicts (Blaikie and Brookfield 1987; Peet, Robbins, and Watts 2011), is a critical theory of social and environmental relations that ultimately "refuses to reduce culture to nature or nature to culture but operates productively in the space between the two—the relationship between signifying and other practices, on the one hand, and an extralinguistic material reality, on the other" (Biersack 2006:28). It is an environmental social theory that "*explicitly* assert[s], as a problem, the inseparable relation between values and the construction of relationships within a world that can always already be deciphered in terms of values and relations" (Stengers 2010:33, emphasis in original). In the spirit of this theoretical angle, this book further develops what I have termed elsewhere the "political ecology of mitigation" (Little 2013a), a new relation to be added to the anthropology and political ecology of technological disaster in late industrial times.[2] I argue that a dynamic and critical political ecology of mitigation attentive to lived experience or "embodied subjectivity" (Merleau-Ponty 2002 [1962]) provides a tool for thinking about residents' experiences of distress, struggle, and negotiation engendered by the IBM pollution conflict. I contend that such a political ecology perspective grounded by ethnographic description shows how toxic substance exposure (e.g., TCE), corporate deindustrialization (e.g., IBM downsizing), and toxics mitigation are socially experienced, articulated, and made meaningful. It is a perspective or angle of thought that favors discursive sorption—that is to say that local discourses on deindustrialization, contamination, and mitigation are *attached*. While I am hesitant to endorse "post" theorizing, as I think more in lines of meshed processes that never clearly break from some temporal constant, the political ecology of mitigation guiding this book is inspired by a post-structural political ecology (Escobar 1996, 1999) that "focuses on the role of language in the construction of social reality; [it] treats language not as a reflection of 'reality' but as constitutive of it" (1996:46). Focusing on the role of discourse to understand the social productions and meanings of the tangle of deindustrialization, contamination, and mitigation dovetails with my effort to expose the narrative materiality of TCE risk, embodied responses to both risk mitigation technologies and the scientific practice and knowledge underpinning contemporary studies of vapor intrusion. "People matter in toxic spills," is a laudable message, and it will be argued throughout this book that another important message is that mitigation politics— a political assemblage where the dynamics and ironies of science, corporate power, and neoliberal governance are exposed and brought to life—also matter in toxic spills.

The ubiquitous presence of TCE contamination and vapor mitigation systems on the Endicott landscape has defined the community, in many ways, a "contaminated community" (Edelstein 2004) in general and a *mitigation landscape* in particular. For the purposes of this book, *mitigation landscape* is a term that "operates both as a password and as a watchword" (Bourdieu 2003:74–75). It is a password to explain a transformed technological, ecological, and biopolitical landscape. As a watchword, it signals the spatial politics of the risk management ethos of public protection against intruding toxic vapors. The term is employed to mean several things at once. First, I use the term to refer to the socio-environmental space within or proximate to the identified boundaries of a known toxic plume that has a vapor mitigation system (also known as a sub-slab depressurization system). Second, a mitigation landscape is a space where corporate and state efforts of toxics response and repair have and continue to take place. In this sense, it is a landscape of *public* technological disaster response where risk decisions and management not only have taken place, but have materialized in the form of vapor mitigation systems. Third, the term marks an occupied territory not just of new mitigation technologies, but of localized subjects living with those technologies of attenuation. And finally, the landscape I refer to is a networked landscape of science and expertise, technology, and forms of assurance. It is an experienced landscape of "actor networks" (Latour 2005), a space where citizens, knowledges, and technologies interact. To describe a *mitigation landscape* is to "crack open" (Murphy 2006) what is unfolding and happening in that landscape. In Endicott's IBM plume zone, a whole lot of science and technology action is at work. This action and its continued moral hold on the community helps define the new mitigation landscape explored here.

Landscape, space, and place have intrigued geographers and anthropologists for some time now (Basso and Feld 1996; Harvey 1985; Low and Lawrence-Zúñiga 2003; Tuan 1974, 1977; Zukin 1991).[3] How landscape, space, and place are experienced has much to do with the process of "emplacement" (Englund 2002; Reno 2011), a phenomenological theory emphasizing how material, experiential, and discursive processes inform place attachment and sense-making (Casey 1996). The concept of emplacement implies that "the subject is inextricably *situated* in a historically and existentially specific condition, defined, for brevity, as a 'place'" (Englund 2002:267). The mitigation landscape explored here can be thought of as a "cognitive landscape" (Latour 2007:27), even an "autogenous atmosphere" (Sloterdijk 2011:46) shaped by entanglements of citizens, toxic substances, sciences, technologies, and the materialization of corporate-state decisions and practices. I propose a landscape or space perspective that welcomes "microsphere analysis" (Sloterdijk 2011) and shares the idea that "Humans have never lived in a direct relationship with 'nature' . . . their existence has always been exclusively in the breathed, divided, torn open and restored space" (Sloterdijk 2011:46). The *mitigation landscape*, in this sense, is not a bare landscape at all. It is a lived, experienced, sensed, and breathed landscape; an "atmospheric-symbolic place" (ibid.) that is seen, heard, smelled, and felt.

We might also think of this tainted late industrial landscape as one representing "a microcosm of social relations" (Zukin 1991:18) or even an *actor network* landscape. Arguing for an "Actor-Network Theory," Bruno Latour proposes that "non-humans" may have an active role, and not be "simply the hapless bearers of symbolic projection" (2005:10). Mitigation is, in this sense, an actant in Endicott's semiotic landscape and a touchstone for exploring the spatialities of life and materiality (Hinchliffe 2007). The mitigation landscape is a resorted environment of quarrelsome subject-object or citizen-technology relations that inform resident negotiations with and translations of life and living in what Isabelle Stengers would call "a modified problematic landscape" (2011:122).

While one might expect confusion, frustration, and suffering to attenuate in zones of mitigation, toxic exposure confusion and uncertainty (Auyero and Swistun 2007, 2008) are social themes that continue to inform place (and home) attachment where risk mitigation has taken place. In this sense, being *situated* in a landscape of toxics mitigation is much like dwelling in a contaminated space of added confusion, uncertainty, and skepticism. Whether occupying a contaminated property or a mitigated property, in other words, the property-liberty relationship is disturbed.[4] The problem of "impaired enjoyment" (Reno 2011:527) endures for residents situated in a mitigation landscape, turning the anthropology of toxics disaster into an anthropology of discomfort and difficult dwelling.[5]

Mitigation systems create commonality on Endicott's plume landscape. They establish a kind of foundation or starting point for a new, shared, and often stigmatized, identity. As described later in more detail, I argue that residents living with TCE contamination and vapor mitigation systems also inhabit a landscape of political ecological processes that influence local experiences with toxics mitigation. Like other political ecology work (Escobar 1998, 1999; Fairhead and Scoones 2005; Moore 1993; Peluso 1992; Sundberg 2003), my focus is on the exposure, illumination, and translation of local understandings of "environment" and addressing contemporary theoretical overlaps between political ecology and science and technology studies (STS) (Goldman, Nadasdy, and Turner 2011). While political ecologists have generally focused their attention on the diverse ways in which "nature," "environment," and "ecology" are perceived differently by different groups (e.g., scientists and laypeople), STS scholars have generally attended to the complex and varied ways in which "science" and "expert" knowledge is produced, applied, circulated, and meshed with political and economic agendas (Hess 1992, 1997b; Hess and Layne 1992). While political ecologists have contested the notion that "environment" and "culture" are cleanly demarcated, STS scholars in general and researchers engaged in the anthropology of science[6] in particular have since at least the 1980s "begun to dismantle the notion that science is carried out in a world scissored off from the culture in which it exists" (Helmreich 1998:19). Therefore, it might come as little surprise that these two fields have "begun to merge, overlap, and borrow from each other" (Goldman et al. 2011:5):

Political ecologists have increasingly sought to better understand the politics of environmental knowledge by reaching out to incorporate the concepts and tools developed in STS. Some STS scholars have likewise sought to engage more seriously with the politics surrounding the application of environmental scientific knowledge by blurring the assumed boundaries between environmental scientific practice (production) and management (application).

To augment the vocabulary of this synthesis of political ecology and STS, the present work highlights the micropolitics of deterritoriality and reterritoriality. These two concepts, which have their roots in the work of Deleuze and Guattari,[7] theoretically ground residents' critical narratives of IBM deindustrialization and risk mitigation efforts. Kuletz (1998) has used the concept of deterritoriality in her study of nuclear testing and contamination in the western United States to theorize "zones of sacrifice": "Once made visible, the zones of sacrifice that comprise these local landscapes can begin to be pieced together to reveal regional, national, and even global patterns of *deterritoriality*—the loss of commitment by modern nation-states (and even the international community) to particular lands or regions" (Kuletz 1998:7; emphasis in original). Of course another major player in this game of deterritoriality is the multinational corporation which banks on its power to settle, accumulate capital, and abandon spaces of production whenever and wherever possible. Put another way, "When the big corporations get rid of their hard properties, complex machinery, invincible walls, heavy voluminous means of production; when they even desert their premises and retain only their logo, the name, flag, colors, sign, and advertisement, they just perpetuate the movement of deterritorialization" (Serres 2011:23). The *mitigation landscape* exposes one such stain of IBM deterritorialization and the technologies at work in this emergent ecology or territory are technologies of symbolic *effort*. In fact, according to the Federal Emergency Management Agency, which in post–hurricane Katrina times has struggled to restore its credibility, "effort" holds symbolic weight in their definition of mitigation. As stated on their government website, "Mitigation is the effort to reduce loss of life and property by lessening the impact of disasters."[8] Mitigation is about creating less risk, not extinguishing risk.

Endicott was and is a zone of IBM sacrifice, a site of corporate exodus. It is a place that was once home to a committed IBM workforce and a place nicknamed "Main Street IBM." Now, three hundred-plus acres of downtown Endicott are marked by a plethora of vapor mitigation systems to manage the risks of widespread TCE contamination. In this sense, IBM deterritorialization has been replaced by another issue, that of vapor intrusion science and mitigation decisions orchestrated by state agencies that have, in effect, *reterritorialized* downtown Endicott. It is deterritorialized, in one sense, because of its post-IBM status and reterritorialized with vapor mitigation systems and remediation efforts that make up the "new Endicott." Surely, there is little that is new about economies or political economic "efforts" contributing to landscape transformation (Cliggett and Pool 2008), and one could easily contend that this "new Endicott" is simply the outcome of capitalism's "restless formation and reformation of geographical landscapes," as described by geographer David Harvey:

> Capitalism perpetually strives . . . to create a social and physical landscape in its own image and requisite to its own needs at a particular point in time, only just as certainly to undermine, disrupt and even destroy that landscape at a later point in time. The inner contradictions of capitalism are expressed through the restless formation and reformation of geographical landscapes. This is the tune to which the historical geography of capitalism must dance without ease. (Harvey 1985:150)

The Rust Belt community of Endicott knows this capitalistic process all too well, a process that might be better translated as "the curse of wealth" (Renfrew 2011). It is experiencing simultaneous IBM deterritoriality and State reterritoriality and undergoing a kind of science and technology facelift. For example, Endicott has shifted from a place dominated by science and technology for high-tech capitalism (with Endicott Interconnect Technologies, Inc. continuing to produce, but certainly not dominating to the degree that IBM did) to a contaminated place marked by state, federal, and private consulting scientists and experts investigating occupational and public health risk, the vapor intrusion pathway, and retrofitting the community with mitigation technologies and experimenting with both standardized (e.g., pump-and-treat) and cutting-edge (e.g., in situ thermal remediation) groundwater remediation techniques. These shifts have social implications in that they potentiate a change or reconfiguration of citizens' views on vapor intrusion management decisions in particular and TCE risk management decisions in general. Deterritoriality and reterritoriality, therefore, are useful concepts from which to navigate the lived effects of deindustrialization, irreversibility, and the politics of repair unfolding in a time of neoliberal governance.

Science, Risk, and Neoliberal Knowledge Politics

Today, a very real problem that residents of contaminated communities face is that contemporary "neoliberal states typically favour the integrity of the financial system and the solvency of financial institutions over the well-being of the population or environmental quality" (Harvey 2005:70–71). These states "tend to favour governance by experts and elites" (67), they are "suspicious of democracy" (66), and even fear special-interest groups and citizen action groups (77). Neoliberal governance also informs contemporary environmental health risk assessments (Stephens 2002), assessments based on technocratic decisions or "state simplifications" (Scott 1998:3) employed to quantify problems, to constrain debate.[9] In other words,

> Although they routinely concede by way of preface that calculation can never replace political judgment, cost-benefit and risk analysts clearly want to rein it in as much as possible. So, typically, they insist that a decision can never be left to the judicious consideration of complex details, but must always be reduced to a sensible, unbiased, decision rule. An effective method should not be a mere language, focusing discussion on central issues, but must be constraining. (Porter 1995:189)

Furthermore, both the EPA and the ATSDR sustain the belief that their science and expert knowledge system is unbiased and therefore accessible to all stakeholders and parties involved in risk debates, even though risk experts and science advisors are in fact trained, shaped, and situated by the practice and culture of science (Pickering 1992; Proctor 1991). These concerns have occupied early anthropological critiques of risk.[10]

Scientific risk assessment is both a powerful tool of the "risk society" (Beck 1992, 1998, 2006; Giddens 1990) and a stark example of the precarious position of science and expertise in contemporary risk society.

This intertwining of the stabilization and destabilization of science and expertise is a dominant characteristic of a world risk society that "provides an especially interesting theoretical backdrop against which to situate an analysis of the relationship of citizens to experts in environmental struggles" (Fischer 2003:48). I am interested here in this same relationship, but shift attention to the anthropological problem of risk that stands out in Endicott: the knotted relations of citizen struggle to risk mitigation technologies and decisions composed and managed by corporate-state science and expertise. The moral hold of risk mitigation in a community disaster context saturated by microelectronic modernization and "technocapitalism" (Suarez-Villa 2009), I shall argue throughout, calls for critical anthropological diagnosis. Assessing risk mitigation, we shall see, is open to the same critiques of risk exposing the hegemonic tendencies of techno-scientific rationality.

As a methodological tool or instrument of the risk society, risk assessment has been understood as a technocratic device that uses science (or rationality) to assist or even counter community understandings of risk (or irrational, exaggerated, and disorderly lay concern). Quantitative risk assessment is an epistemic decision that reinforces techno-scientific rationality and the authority of the state, with the ultimate goal of supplying "the public with 'objective' information about the levels of risks. That is, the 'irrationality' of contemporary political arguments is to be countered with 'rationally' demonstrable scientific data" (Fischer 2003:127). Fischer adds that this scientifization of risk aims "to offset the misperceptions and distorted understandings plaguing uninformed thinkers, particularly the proverbial 'man on the street.' The objective of this new line of research has been to figure out how to more persuasively convey the technical data to override the 'irrational fears' of ordinary citizens" (2003:127). Others have argued that the social meanings of risk need to be rectified, made an object of social scientific study, and brought into public environmental debates if we truly wish to have an effective dialogue between industries and technologies producing risk and communities struggling with toxic contamination (Wynne 1989). In order to avoid "common misconception[s] about the nature of scientific rationality (and about public 'irrationality') embedded in our risk management institutions" (35), social scientists have turned to the study of science or science culture.[11]

Anthropologists, sociologist, historians, philosophers, and other scholars working within the broad field of STS have shown that science and the knowledge it produces is anchored by social, cultural, political, and economic context (Franklin 1995; Harding 1991a, 1991b; Jasanoff 2004; Knorr-Cetina 1999; Latour 1987; Latour and Woolgar 1979; Nader 1996; Rajan 2006, 2012). These critical studies offer perspectives and theories of knowledge production that can easily apply to critical scientific risk assessment debates. One basic contribution of these science studies is the perspective that science is a social practice, a meaningful social phenomenon open to ethnographic investigation. From this perspective, risk assessment, like all scientific information, is about the slow tactical development of data and facts, about science *formation* or science-in-the-making (Latour 1987). The scientific facts and data employed to formulate a risk assessment report are "not only collectively transmitted from one actor to the next, [they are] collectively *composed* by actors" (Latour 1987:104, emphasis in original). In this way, scientific facts get their power from inter-subjective practice, socialization, and collectivization. Facticity in general and risk ontology in particular is the outcome of a social and political economic reality conditioned by production, construction, and competing forms of representation. These dynamics of scientific knowledge production are worthy of ethnographic investigation, especially since science has long been and continues to become a strong culturally and morally transmitted value in the contemporary world (Shapin 2008).

The view of science in the closet or science as an epistemic practice with ready-made facts to work with is missing the actual *practice* of science (Latour 1987; Latour and Woolgar 1979). In this way, the anthropology of scientific risk assessment ought to think of risk in terms of form, as formed fact, in the way that Latour and Woolgar (1979:28–29) have encouraged us to appreciate "how science is done" and how facts are constructed and made. Risk or the value of risk is not singular or self-evident, but rather emergent and given form. Furthermore, risk ontology is politically charged in the same way that knowledge and the state are "co-produced" (Jasanoff 2004). Politics is not "something to be *added on* to risks after they emerge either from nature or social interaction, but at their core risks are essentially political phenomena" (Hiskes 1998).

Another useful perspective from these critical social studies of science is that scientific knowledge (e.g., scientific risk assessment data) offers only a "partial perspective" (Haraway 1991) in the same way that local understandings of risk are partial perspectives. Haraway (1991) employs a science studies perspective that insists "on the embodied nature of all vision" and the strategic avoidance of accepting theories of knowledge that claim to be a "gaze from nowhere" that somehow perfects our knowledge of universality (188). It is precisely sloppy critiques of science that Haraway seeks to avoid. Instead, science, like all ardent cultural epistemologies, has to be argued *for* in a new critical way. In her words: "Science has been utopian and visionary from the start; that is one reason 'we' need it" (1991:192). In a similar vein, Laura Nader (1996) argues that the ultimate goal of anthropologies of science "is not to 'put, scientists in their place' (although one might want to). . . . The point is to open up people's minds to other ways of looking and questioning to change attitudes about knowledge, to reframe the organization of science" (Nader 1996:23).

Haraway (1991:193) reminds us that *connection* is a real aim of objective thinking: "a scientific knower seeks the subject position not of identity, but of objectivity; that is, partial connection." In other words, if scientists were to really be working with universality or "totalization" they would not be in a position to get to know the subjects of science—data, microbes, or the spoken word. Getting to know the subjects of risk—the toxin, the experimental rodent, the human—is, or ought to be, according to Haraway (1991, 1997), the goal of empirical scientific studies of risk. "Partial connection" creates knowledge and understanding.[12] Citizens and scientists *connect* to risk debates with equal force, up until the point at which the optics of politics are focused and the demarcation of lay and expert becomes clear. As Douglas and Wildavsky (1982:174) have reminded us, all risk debates emerge from or end up igniting a "political dialogue" with inequality and power relations. In addition, risk assessment specialists and other public health scientists tend to disconnect and go to the lab to study samples and interpret results, while residents living in contaminated communities stay connected.

Other scholars of the society-science interface have explored the public understanding of science, showing that science and the workings of science often rank low on the public's list of interests (Irwin and Wynne 1996). The scientific study of risk and especially the complexities of environmental health risk analysis may in fact *not* be a matter of concern or interest for the lay public. As Irwin, Dale, and Smith (1996) observe:

> from the perspective of the scientific community, science "disappears" within everyday life; from the perspective of local people, science has no necessarily special or privileged status alongside such routine concerns as unemployment, rent increases, [property devaluation], or factory pollution. Issues such as "knowledge" or "hazard" thus form part of a wider pattern of everyday social relations. (52)

Irwin and Wynne are not arguing that the public has an *essential* tendency to misunderstand science. Instead, they explore how and perhaps why citizens within specific situations and contexts can and often do misunderstand science and its ability to make sense of the world, at the same time that science *misunderstands* "both the public and itself" (Irwin and Wynne 1996:10). What is more likely occurring in society is a situation in which scientific experience and knowledge and citizen experience and knowledge are simultaneously co-produced. Additionally, scientific ambiguity and citizen experiences with ambiguity are also being co-produced. Perhaps most unique about these issues from a "risk society" (Beck [1992] 1986) perspective is that these struggles with uncertainty have emerged and continue to survive in a late modern world swarming with high technology, advances in scientific complexity, and a growth industry of information, knowledge, and scientific expertise. In other words, "modern" institutions, values, and technoscience are amplified as a result of growing natural and technological disasters. These logics of modernity don't retract in the situations of risk response. For Beck, "[t]his is what 'reflexive modernization' means: we are not living in a *post*-modern world, but in a *more*-modern" (2006:338).[13] What matters, according to Beck, and what seems to matter in contaminated communities like Endicott, is that "Risk definition, essentially, is a power game. This is especially true for world risk society where Western governments or powerful economic actors [like IBM] define risks for others" (2006:333).

Anthropological studies engaging the sociopolitics of toxics risk, which usually involves sociopolitical discussion of science and expertise, explore the diversity of social actors making up the risk decision-making environment (Barker 2004; Button 2010; Fortun 2001; Petryna 2002). These anthropologies of toxic contamination are challenged by a core research question: "How does one account for the way disaster [or toxic contamination or mitigation] *creates* community?" (Fortun 2001:10; emphasis in original). "Struggles over what will count as rational accounts of the world are struggles over *how* to see" (Haraway 1991:194; emphasis in original). Anthropologists studying risk and toxic contamination debates are interested in *how* people see the contaminated environment, their relationship to it and that of others, and what they draw on to understand and make sense of their situation and how to improve their situation.

These qualitative studies draw attention to social and political relations and the prudent knowledge politics that emerge from toxic contamination, treating the risk analysis debates of scientists and lay struggles to understand as equally worthy of, and in need of, ethnographic investigation. Furthermore, "even when quantitative data are needed to determine the existence of environmental health effects, qualitative data are necessary to understand how people and communities experience and act on these problems, as quantitative data can only render an imperfect or partial picture of health effects and their causes" (Brown 2003:1789). The very inconclusive nature of formal scientific risk assessments can be thought of as a gap or inroad through which anthropology and other social sciences can be applied.

Disaster and Narration

Knowledge and uncertainty, "perceptibility and imperceptibility" (Murphy 2006), are simultaneously produced in spaces of technological disaster. Within this terrain of toxic intrusion, the power of the corporate state is often reinforced, ironically, by the very absence of concrete and "official" scientific knowledge of risk. In other words, within a "climate of scientific uncertainty, corporations cast doubt on their critics and seek to undermine challenges to the *official* narrative and attempt to avoid the scrutiny of government agencies" (Button 2010:13; emphasis in original). In addition to doubting *unofficial* counter narratives, "corporations often resort to defining the [toxic] event in a more limited scope in order to prevent government interference. Often they dismiss claims of critics by labeling them a junk science, 'irrational', or 'overly emotional' (2010:13).[14] This remains a central feature in the anthropology of disaster writ large, as more often than not "science cannot address the asymmetrical balance of power that lurks in the wings of all disasters. The use of power to amplify uncertainty is a perfect example of how official narratives constrain and overpower alternative narratives" (248).

None of this is meant to drag science through the mud or discredit the epistemic power of scientific reasoning or the value of sustaining a scientific and technological society. I am not anti-science. Instead, toxic disasters, like the one depicted in this book, showcase how corporate and state responses to technological disaster result in practices of risk control and mitigation in socio-environmental contexts that at are actually extremely messy and out of control. As Kim Fortun's study of the Bhopal disaster highlights so vividly: "Disaster is not confronted as disaster" (Fortun 2001:52). Instead, disasters—from Methyl isocyanate contamination from a Union Carbide explosion in Bhopal, India, to IBM's toxic spill in Endicott—and the risks they pose are reified, isolated, and reduced as problems to be managed, remediated, and mitigated by experts and their actions controlled and supported by quantification and fancy computational modeling. In this way, TCE mitigability and controllability reinforce the position of techno-scientific authority and state power, even when state regulators and scientists themselves question the possibility of 100 percent TCE remediation, or total "control." Again, pointing to the socio-political nature of vapor intrusion science and the state's role in shaping mitigation science and policy is not about debunking these practices and ways of knowing or suggesting "official knowledge" is devoid of "internal fissures and tensions" (Mathews 2011:14). Instead, "We need the sciences now more than ever, but we have to have them reimagined and re-formed—reformed by attention to their own history, by overt attempts to enact them differently, and by new protocols for questioning and judging the sciences as they develop" (Fortun and Bernstein 1998:xii).

Anthropologies of disaster, toxic contamination, and risk debates make this messiness more audible and expose the various concerns of the mix of social actors. As Fortun (2001) instructs, "The challenge is not to stand outside established systems [i.e., systems of risk assessment science and expertise], but to find places to work within them—finding repose in none, recognizing that neither poetry nor law can tell the whole story" (37). More recently, Kim Fortun and Mike Fortun (2005) rightfully contended that anthropological critiques of environmental health expertise need to go beyond simply challenging the undemocratic practice of experts or the "tendency of experts to silence lay voices, although they often do. . . . Anthropological works that engage the sciences as domains in which ethics are worked out have more promise. . . . Anthropology can itself become an ethical plateau that supports and presses on toxicologists" (Fortun and Fortun 2005:50). As this book argues, as a discipline focused on human experience, anthropology has equal promise in sites of vapor intrusion risk, in sites of unsettled technological disaster.

What amounts to a study of "spatial stories" (de Certeau 1984) of deindustrialization, contamination, and risk mitigation response, I attend here to the particular ways in which narrations of IBM's birthplace disaster zone are "activated and amplified within narrative communities" (Allen 2003:22). This "narrative community" perspective is very different from the modernist version of human agency whereby a so-called essential will to speak is what makes individuals speak out and expose their identity. Similar to Fortun's (2001) "enunciatory communities" approach, the "narrative community" perspective insists that "identities come together by sharing experiences, creating new networks of stories and constructing an alternative vision of an unjust present and a promising future in toxic communities" (Allen 2003:23). Following de Certeau's (1984:118) perspective that "Stories . . . carry out a labor that constantly transforms places into spaces or spaces into place," one could argue that narrating the mitigation landscape is about making *spaces* of mitigation into *places* of mitigation where the actualization of mitigation is experienced, practiced, and articulated.

Certainly, bringing such critical narrative perspectives on risk and mitigation into the environmental public health sector in general and vapor intrusion sciences in particular involves confronting a very basic crux: "It is well and good to commit to the belief that science always occurs in a cultural context; its conceptual categories, its rule of evidence, its distinction between appropriate and inappropriate subjects for investigation all reflect the society within which scientists work. But constructivism [and discourse-based political ecology for that matter] has never made a big splash in government" (Tesh 2000:101). Despite recent attempts to democratize vapor intrusion governance by increasing community "stakeholder" participation (see chapter 7), the fact is that ethnographic narrative is devalued in this terrain of techno-science, if not totally ignored, even amid growing partnerships between techno-scientific experts and activists involved in grassroots movements (Ottinger 2013; Ottinger and Cohen 2011). Following Allen (2003), I believe these narratives ought to be liberated from the so-called bullying of quantification: "residents' narratives form part of a liberatory civic discourse that facilitates social and environmental change. . . . By focusing on residents' verbal testimony, narrative strategies are acknowledged as a way of knowing about one's environment that is different from corporate or scientific descriptions" (Allen 2003:19). A turn to local narratives of socio-ecological struggle is not about ousting the value of quantitative scientific risk assessment; rather, it opens up another relation in the "ecology of practices" (Stengers 2011), another epistemic opportunity or point of reference that can be drawn on to formulate an alternative environmental health perspective and disaster narrative, and one that is much needed to learn about dynamics of vapor intrusion debates and politics unfolding in troubled communities like Endicott.

In sum, in an attempt to articulate the complex relations of deterritoriality and reterritoriality, risk mitigation politics, and the paradoxes of corporate responsibility in late industrialism, I entwine political ecology and STS insights to develop a nuanced environmental anthropology of high-tech disaster and toxics repair. As a political ecology study of multiple layers and flavors, the book uses the *mitigation landscape* as a conceptual anchor to rethink narratives of IBM deindustrialization and community corrosion, as well as to critically diagnose corporate and state techniques of "responsible" pollution action. What follows, then, is a political ecology of troubled life in a "mitigated" technological disaster zone. Along the way, we see how risk mitigation efforts are not simply a techno-scientific response to technological disaster, but processes involving breathing and thinking people, both lay and expert. We also learn about IBM's birthplace residents and their lived ecologies. We witness their entanglements or "networked" (Latour 2005) experience with science, technology, corporate-state power, and advocacy politics, and their confrontations with new transformations of space, place, and landscape in late industrial times.

3

From Shoes to Computers to Vapor Mitigation Systems

I think someone said that Endicott is ground zero for TCE
contamination or something, but I don't know. But I'll tell
ya, it is definitely where IBM started. Everyone here knows
that.
—Endicott resident and activist, 2009

We witness a transformation of substances and a dissolution
of forms . . .
—Deleuze and Guattari (1987:109)

Making sure I understood things properly, that I understood the significance of Endicott's industrial history, I was told by one "plume resident" during an interview: "You know, Peter, Endicott is where IBM
started? It all started right here." I was reminded of this fact on several occasions, and no matter how many times I heard it, I still found it
hard to believe that Endicott, a small village in western New York, was
not only IBM's birthplace, but also the place where some of the earliest computing technologies emerged to help create the "third industrial
revolution" (McGraw 1997) and the so-called Information Age.

But before IBM, there was the Endicott-Johnson Shoe Company (EJ),
and the charismatic force of EJ's founder, George Francis Johnson (or
'Georg F' as locals refer to him). EJ, with the leadership of George F,
turned Broome County into the "Valley of Opportunity," and Endicott
was often called "the Magic City" because of the speed at which it shifted
from a landscape of farmlands and swamps to a vibrant and profitable

commercial and manufacturing center. Both EJ and IBM played critical roles in the area's industrial facelift. As one long-time resident put it, "Endicott wanted to be the biggest thing between New York City and Buffalo," and EJ and IBM made this industrial utopian desire a reality.

Most residents I interviewed referred to Endicott as a "two company town" that was booming until EJ moved in the late 1960s and IBM began downsizing in the 1980s and '90s, eventually closing its Endicott facility in 2002. Today, Washington Avenue, the consumer and social epicenter of downtown Endicott since the early 1900s, has bottomed out. "Today you could shoot a cannon ball down Washington Ave and hardly hit anyone," were the chosen words of one elderly resident. This scenario highlights the basic narrative of Rust Belt communities: industries develop, industries leave to cut costs, and people suffer the consequences as a result of loss of jobs, decline in local business, pollution and health risks, and property devaluation. Now most residents think of Endicott as an industrial wasteland with a dissolved tax base, or, as one agitated resident proposed, "Endicott should be renamed Emptycott. There's nothing here anymore, except this stupid plume." How this shift occurred will be discussed in more detail later in the chapter.

After a brief discussion of EJ history, I focus on the emergence of IBM, its tangled history of company mergers, and how Endicott figures in the history of one of the computer industry's most powerful and influential corporations. As McGraw points out, "The story of IBM is an epic in the business history of the United States. Through it run some of the main currents of twentieth-century American life: the Great Depression, the administration of the welfare state, World War II, the antitrust movement, and the Cold War" (McGraw 1997:349). Moreover, IBM helped "create" modern capitalism by developing the earliest machines used to send memos around the world, expedite banking transactions, and make public databases electronically accessible. IBM played a dominant role in transforming information management and business practice by replacing pencils, pens, and ledgers with adding machines, cash registers, and manual typewriters designed to modernize industry.[1] IBM helped plant the electronic seeds for "high-technology capitalism" (Dyer-Witherford 1999), "digital capitalism" (Schiller 1999), "technocapitalism" (Suarez-Villa 2009), and the broader computerization of culture, communication, and human experience.[2] This chapter focuses on some of the products, the machines, developed at the IBM-Endicott factory and their critical role in the modernization of computing technology.

After the midpoint of the twentieth century, Endicott, like many other Rust Belt communities of the Northeast and Midwest, shifted from industrial powerhouses to places suffering the symptoms of deindustrialization, dissolving tax bases, the rise and expansion of the service and information economy, the suburbanization of capital and people, and the arrival of a new "global" neoliberal economy. What follows, then, is an historical account of "high tech in the Rust Belt" (Grossman 2006:99) that focuses on Endicott's industrial and late industrial narrative as it relates to broader themes of Rust Belt reality, including deindustrialization, corporate dynamism, and "creative destruction"[3] (e.g., corporate downsizing, cheap labor adventurism, etc.). Included in this discussion, is a brief overview of the current high-tech employer to replace IBM, Endicott Interconnect Technologies, Inc. (EIT). While EIT is staying afloat in the highly competitive high-tech market, residents realize that it will never become the employer that IBM was.

Reminding me of Endicott's Rust Belt status and the dystopia invoked by both IBM's closure and the toxic spill, one resident told me "IBM cut this community's throat." Another goal of this chapter, then, is to expose the toxic leftover of this so-called IBM homicide or what my more modest residents describe as normal corporate "abandonment" or "business-as-usual downsizing." To do this, I explain the messy history of the IBM-Endicott chemical spill, including the history of the groundwater remediation effort at the site, the government agency response, the emergence of the vapor intrusion risk and mitigation, and, finally, the development of local grassroots advocacy.

Endicott-Johnson and the Lure of George F

Any discussion of industrial history in Endicott and the greater Broome County area must begin with the widespread influence of the Endicott-Johnson Shoe Company (EJ). Chances are that anyone who grew up in Endicott had either a father, grandfather, or other relative who worked at one of EJ's factories. Incorporated in 1906, EJ was a prosperous manufacturer of shoes based in New York's Southern Tier region, with factories mostly located in the "Triple Cities," the name for the triad manufacturing influence of Binghamton, Johnson City, and Endicott.

Local residents, especially "old timers," tend to admire George F and EJ's role in Endicott and the regions' industrial history, despite common curiosity regarding the environmental impact of EJ's tannery operations. Here are several residents' reflections on EJ and George F:

> My dad worked for EJ for forty five years. EJ did a lot for the community. . . . There was never any question that when Endicott Johnson put on some event, that everyone in the community was invited and involved. [Retired banker]

> My dad worked at EJ. Back then, years ago, Endicott-Johnson in this area was huge. At one time, I think just after WWI, Endicott-Johnson was the second largest shoe industry in the United States. From what I have read, during the mid to late 1940s, EJ was producing over 50 million shoes a year. [Retired welder]

> George F. built churches, he built parks, and he built lots of homes for people who had bad luck. He would tell them to just pay the interest. They didn't have to pay the whole amount. My husband's father lived in an EJ home and it was bought for $3,300, brand new. I'll tell ya, George F. did more for Endicott than IBM ever did. [Retired receptionist]

By the 1920s, an estimated 20,000 people worked in EJ's factories, with the labor population booming during EJ's golden years in the mid-1940s. This growth spurt was a direct outcome of World War II, since EJ produced an abundance of military footwear during the war years. During the early 1950s, the work force still numbered approximately 17,000 to 18,000. Although EJ was established by two men, Henry B. Endicott and George F. Johnson, the latter really was *the* laudatory figure marking the industrial landscape of the region. Worker respect for George F is symbolized by two arches, one in Johnson City and the other in Endicott, which were erected by EJ workers in the 1920s to honor his "Square Deal" philosophy that workers deserve "a fair day's work for a fair day's wage." This deal was fueled by his belief that "Capital is an empty, brainless thing and couldn't live a minute without labor" (Inglis 1935).

Motivated by the general philosophy of welfare capitalism and other Progressive Era movements of the early twentieth century, George F advocated for the development of parks, churches, libraries, and carousels[4] to uplift and empower workers and their families. The deal consisted of worker benefits—even amid harsh economic times—that were generous and innovative for their time, including building homes and offering loans to workers who could not afford down payments, as alluded to in the resident's quote above. This commitment was for him a matter of viewing the home as a symbol of "security": "There can be no security—there can be no guarantee of prosperity and industrial space—except through homes owned by the plain citizens. I believe myself that the home is the answer to Bolshevism, Radicalism, Socialism, and all other 'isms'. You will find that the home is the basis of all security." These actions aimed to augment worker loyalty, but they were also designed to discourage unionizing, or at least level its legitimacy. George F is remembered for being a father of "industrial democracy" (Inglis 1935), even an "Industrial King," the title of one of the songs played by EJ's famous marching band. EJ workers showed their pride in EJ and George F by singing songs with the lyrics:

> In Endicott there lives a man, Whose fame is far and wide;
> For making shoes he leads them all, For kindness, too, beside;
> The 13,000 under him Look up to him as guide,
> And swear by the man Who treats them fair.

Another EJ song compared Endicott to "Paradise," a community where Heaven was imported by George F:

> You say there's no dissension and that trouble is unknown
> Then you folks must be living in a small world of your own;
> What's that? No strikes or agitators screeching out their lies?
> You say the name is Endicott; I'd call it Paradise.
> You shake my hand and mean it in this little spotless town,
> And every smiling face is just a stranger to a frown;
> George F. can't get you into Heaven, but I tell you what,
> He did as much when he brought Heaven here to Endicott.

George F's paternalistic drive inspired the business philosophy of Thomas J. Watson, Sr., who would become the founding father of International Business Machines Corporation (IBM) and a major player in the development of what today is termed "technocapitalism" (Suarez-Villa 2009). While contending that "paternalistic companies are a thing of the past," one retired IBMer I interviewed informed me that "IBM and EJ were what you would call paternalistic companies . . . Mr. Watson and George F. learned a lot from one another. Watson listened to George F. and IBM adopted his paternalistic approach."

After EJ's postwar peak in the late 1940s, the company steadily lost market share to domestic and especially foreign competitors from Europe and South America who began to control the shoe and leather industries. EJ's founding principles of valuing labor were threatened by the availability of cheap foreign labor and the rise of global shoemakers like Nike. The company responded to these economic changes by cutting back production and labor, an ongoing process that eventually led to EJ falling behind IBM as the region's industrial power. As the 1950s passed into the 1970s, and EJ's style of shoes fell out of fashion,[5] the Johnson family withdrew from active company management to be replaced by non-family professional managers. EJ's tannery operations were closed in 1968 in response to these economic conditions, and the plan to close the last of its shoe manufacturing plants in the Triple Cities area was not announced until 1998.

The Emergence of Big Blue

International Business Machines Corporation (IBM),[6] the world's largest multinational computer and information technology (IT) consulting corporation, is informed by three core business goals and values: "Dedication to every client's success"; "Innovation that matters – for our company and for the world"; and "Trust and personal responsibility in all relationships." IBM manufactures and sells computer hardware and software, in addition to offering infrastructure services, hosting services, and consulting services in areas ranging from mainframe computers (computers used mainly by large organizations for critical applications including, but not limited to, bulk data processing such as census, industry and consumer statistics, and financial transaction processing) to nanotechnology to genographics[7] to cutting-edge epidemiologic software.[8] Moreover, IBM's Fortune 500 status remains strong despite IBM's departure from personal computer (PC) markets. For example, IBM's 2008 revenues totaled $103.6 billion, nearly twice that of Microsoft's earnings,[9] and it continues to attract frontline investors. In November 2011, for example, billionaire Warren Buffett bought IBM's shares, worth nearly $11 billion, and he now owns 5.5 percent of the company.

Of course, IBM would not be a champion in the IT industry without having an industrial birthplace. While Endicott has gained a certain status as the corporate birthplace and location of IBM's first manufacturing plant in 1906, the actual origins of IBM are more complicated, as is the case with all histories. What follows is a brief history of Big Blue's emergence and the individuals and companies that merged to become one of the most powerful American corporations. Moreover, while the investigative journalist Elizabeth Grossman (2006) accounts for the history of IBM in Endicott in her popular book *High Tech Trash: Digital Devices, Hidden Toxics, and Human Health*—a book, I should add, that was suggested to me by several residents and activists I interviewed—her finding that "IBM began manufacturing business machines and their components in the 1930s" (2006:99) needs greater historical contextualization.

According to IBM's own corporate archive, the earliest roots of what eventually became known as IBM can be traced back to the period 1880–1890. In 1885, Julius E. Pitrat of Gallipolis, Ohio, patented a new device that he called a computing scale, an invention that would become the primary technology and product of the Dayton Scale Division of IBM. At the same time that Pitrat was working on his computing scale, Dr. Herman Hollerith, a distinguished statistician who had been employed by the U.S. government to automate and compile data for the 1890 U.S. Census, was struggling in Washington, DC, with the problem of reducing the mountain of data gathered by census-takers into manageable or usable form. Solution driven, Hollerith set out to find a way to mechanize the recording, tabulation, and analysis of data, information, and facts, and the system he devised was fundamentally simple. It consisted essentially of a method of recording the facts of any given situation—for example, the description of one person—by punching a definite pattern of holes in a piece of paper.[10] His punched card tabulating machine resulted in an expansion of Hollerith's clientele. His first customer was the City of Baltimore, and other early users of Hollerith's machines were the Bureau of Vital Statistics in New Jersey and the Board of Health in New York City.

Developed in 1890, Hollerith's punched card tabulating machine could automate tabulation at a speed of 10 cards a minute. IBM's System/360 was invented in 1964, and could read up to 1,000 cards a minute while performing high-speed computations. The launch of the IBM System/360 was, according to IBM, a significant event in the history of computing. For IBM, it was a staggering undertaking. The company spent three-quarters of a billion dollars just on engineering, and invested another $4.5 billion on factories, equipment, and the rental machines themselves. IBM hired more than 60,000 new employees and opened five major new plants. Thomas J. Watson, Jr., called it "the biggest privately financed commercial project ever undertaken." The timing of the launch, when not all of the new machines had yet undergone rigorous testing, was, said Mr. Watson, "the biggest, riskiest decision I ever made." And when he unveiled the new System/360 on April 7, 1964, he presented it as "the most important product announcement in company history" (IBM Corporate Archive, 9215FQ14, p. 79).

In this way, the System/360 was a significant accomplishment in computer modernization, a modernization marked by speed, tabulation, and computing technologies enabling "business" modernization, innovation, and progress. IBM's System/360 was faster, and faster was better—a fact of computing innovation that continues today, as with IBM's recent celebration of meeting its goal of breaking the "petaflop" or the ability to calculate 1,000-trillion operations every second.[11]

With Pitrat in Ohio and Hollerith in Washington, there were two other figures that played into IBM's development: William L. Bundy and Alexander Dey. Committed to time machines—for time recording and time management—in 1888, Bundy, a jeweler in Auburn, New York, devised a mechanism by which workers used individual keys that were inserted into a time-recorder to log arrival and departure times at work. That same year, Dr. Alexander Dey patented a time recorder that avoided the use of separate keys by allowing workers to swing a pointer on the machine—also known as the Day Time Recorder—to their own employee number. They would then push the punch in the corresponding hole, thereby causing the time record to be printed opposite the employee number on a prepared sheet inside the machine. These early innovations and the following developments led to commercial organizations which later merged to become IBM. These merging companies included the Bundy Manufacturing Company (the first time-recorder company in the world), the Computing Scale Company (manufacturer of the world's first computing scales), the Tabulating Machine Company (incorporated by Hollerith in 1896), the Dey Patents Company (which changed its name to the Dey Time Register Company), the Willard and Frick Manufacturing Company (developed to market the world's first card time-recorder patented by Daniel M. Cooper in 1894), the Stimson Computing Scale Company (a company from Detroit founded in 1896), the Moneyweight Scale Company (founded in 1899), and the International Time Recording Company (ITR) (organized in 1900 by George W. Fairchild to take over the properties of the Bundy Manufacturing Company and its newly acquired subsidiary, the Standard Time Stamp Company, and the Willard and Frick Manufacturing Company). In 1906, having outgrown Bundy's plant in Binghamton, New York, the company began work on a new factory building in Endicott, under the name of International Time Recording Company (ITR). ITR became one of two other seed companies that eventually merged to become IBM.

In 1907, ITR purchased the Dey Time Register Company, and transferred the manufacture of dial recorders to Endicott. The continued progress of the Computing Scale Company of America, the International Time Recording Company, and the Tabulating Machine Company attracted the attention of Charles R. Flint, a businessman and financier. He became convinced that a merger of the three companies would result in a formidable business institution. Around 1910, he approached the three companies with merger proposals, and on June 16, 1911, the Computing-Tabulating-Recording Company (CTR) was incorporated as a holding company to assume ownership of the three parent concerns. To combine them into a single harmonious unit, CTR's board sought out an executive not previously identified with any of the parent companies. That qualified leader was Thomas J. Watson, Sr. In 1914, Watson was hired as CTR's general manager.

It wasn't until February 1924 that Watson, Sr., decided to change the company's name to the International Business Machines Corporation (IBM), to accurately reflect the global reach of CTR's products and services, whose international connections had been well established by CTR's predecessor and partner, ITR. During the 1920s and 1930s,

Watson, Sr. (or T.J. as he was known) shifted IBM's focus away from low-tech machines to more sophisticated technologies surfacing during that time, like the punchcard machine. He believed wholeheartedly that "Machines can do the routine work. People shouldn't have to do that kind of work" (in McGraw 1997:358), and by the 1940s, IBM held 90 percent of the U.S. market for punchcard machines (1997:360).

The IT revolution and the computer era really exploded in the 1950s, when governments, industry, and academics were conducting research using datasets that were impractical or impossible to manage without computing technologies (Jonscher 1994). Equally important was the fact that number-crunching costs declined during this time. For example, between 1950 and 1980, the cost per MIP (million instructions per second) fell 27–50 percent annually, increasing the widespread use of calculating devices. By the 1960s, computers became a standard technology used by businesses, industry, and researchers to sort, store, process, and retrieve large volumes of data. This shift, of course, saved businesses and researchers on the labor involved in information processing activities. It has been suggested that the cost of storage probably fell at an annual rate of 25–30 percent from 1960 to 1985, and the use of computers as communication devices, which began in the 1970s and continues today, has reduced the costs of business travel (Greenwood 1997). Furthermore, the increase in computer-assisted research impacted many fields of science, including the computerization of anthropology itself (see Burton 1970; Fischer 1994; Hymes 1965).[12]

While Endicott's international business connections had already been forged by EJ by the early 1900s, the international or global roots of IBM were planted before the company was named IBM in 1924. In fact, before CTR and ITR officially merged to become IBM in 1924, both had offices and manufacturing facilities already established in a number of countries, including Germany, England, Spain, Portugal, France, Italy, Hungary, Belgium, the Netherlands, Denmark, Scotland, Finland, Norway, Sweden, Switzerland, Argentina, Brazil, Uruguay, Puerto Rico, Canada, and Australia. The notion that "the sun never sets on IBM" (Foy 1975) could not be farther from the truth, especially when considering its early and contemporary global presence.

The global and expansionist philosophy of IBM is embedded in lyrics of numerous IBM songs that were sung at banquets and luxurious dinners at the IBM Country Club in Endicott. These company songs—"IBM One

Hundred Percent Club," "Ever Onward," March On With IBM," and "Hail To the IBM"—mixed pride with profit, corporate spirit with capitalist enterprise. Here is a verse from "Hail To the IBM":

> Lift up our proud and loyal voices,
> Sing out in accents strong and true,
> With hearts and hands to you devoted,
> And inspiration ever new;
> Your ties of friendship cannot sever,
> Your glory time will never stem,
> We will toast a name that lives forever,
> Hail to the I.B.M.
>
> Our voices swell in admiration;
> Of T. J. Watson proudly sing;
> He'll ever be our inspiration,
> To him our voices loudly ring;
> The I.B.M. will sing the praises,
> Of him who brought us world acclaim,
> As the volume of our chorus raises,
> Hail to his honored name.

The lyrics of IBM's rally song "Ever Onward" hint at the culture of globalization established by IBM and recycled by the current neoliberal mantra of uninterrupted economic growth and capital accumulation:

> There's a feeling everywhere of
> Bigger things in store
> Of new horizons coming into view.
> Our aim is clear, To make each year
> Exceed the one before
>
> Staying in the lead of everything we do.
> The will to win is built right in.
> It will not be denied
> And we will go ahead
> We know by working side by side

CHORUS:
Ever Onward, Ever Onward
That's the spirit that has brought us fame.
We're big but bigger we will be.
We can't fail for all can see that
To serve humanity has been our aim.
Our products now are known in every zone.
Our reputation sparkles like a gem.
We've fought our way thru and now
Fields we're sure to conquer too.
For the Ever Onward IBM

Ever Onward, Ever Onward
We're bound for the top to never fall.
Right here and now we thankfully
Pledge sincerest loyalty to
The corporation that's the best of all.
Our leaders we revere and
While we're here let's show the
World just what we think of them!
So let us sing men, sing men, once
Or twice then sing again.
For the Ever Onward IBM

The IBM Endicott facility was time and again celebrated for its breaking records in punchcard production "speed." In 1937, IBM had 32 presses operating in its Endicott plant. These machines were printing, cutting, and stacking 5–10 million punched cards each day—numbers unmatched by any other company using tabulation technologies at the time. Watson, Sr., had a philosophy that informed these innovation breakthroughs: "THINK."

Watson, Sr.. emphasized research and engineering, and introduced into the company his famous motto "THINK." The motto dates back to December 1911, when future IBM Chairman Thomas J. Watson, Sr., managed the sales and advertising departments of the National Cash Register Company (NCR). At a sales meeting, he is reported to have said: "The trouble with every one of us is that we don't think enough.

We don't get paid for working with our feet; we get paid for working with our heads. . . . Thought has been the father of every advance since time began. . . . 'I didn't think' has cost the world millions of dollars." Frustrated by this cognitive deficit, it is reported that Watson wrote "T-H-I-N-K" on the easel behind him. Following this enlightening event, "THINK" was placed on signs in every department at NCR. The one-word slogan traveled with Watson when he later joined the CTR— the forerunner of today's IBM—as general manager in 1914. In the early 1930s—thanks to the process of plastic lamination that facilitated large-scale production and distribution—the THINK motto began to take precedence over other slogans in IBM. It soon appeared in wood, stone, and bronze, and was published in company newspapers, magazines, calendars, photographs, medallions—even *New Yorker* cartoons—and it remains today the name of IBM's employee publication. One can also find THINK etched on buildings at the former Endicott facility. THINK is etched on what locals call the "old engineering building," a facility completed in 1933, and it has been said that by 1940 this modern research and development (R&D) laboratory housed "more scientific ability per square foot than any other building in America" and became an instant symbol of industrial progress (Aswad and Meredith 2005:87). Today, this and other buildings engraved with IBM's THINK motto are surrounded by groundwater monitoring wells to test for TCE vapors.[13]

In the early years (1906–1914), the Endicott site consisted of five buildings, and employed fewer than 300 people. By the time CTR changed its name to IBM in 1924, Endicott employment had tripled. By the mid-1930s, there were 25 buildings at the IBM-Endicott complex. In later decades, the site employed thousands of workers, naming the IBM-Endicott facility "Main Street IBM." By 1939, there were approximately 3,800 IBM employees in Endicott, making up about a third of IBM's total work force at the time. Just days before the start of the 1939 World's Fair in New York City, the entire work force assembled for a photograph to be displayed at IBM's exhibit at the fair, which displayed, among other IBM products, the successful IBM 405 tabulator (originally called the alphabetic bookkeeping and accounting machine, and later known as the 405 electric punched card accounting machine).

The IBM 405 was a combined adding, subtracting, and printing machine that printed complete reports from punched accounting

Figure 3.1 IBM-Endicott Plant, Employee Portrait, 1934. IBM Corporate Archive.

machine cards. It could be used to list both alphabetical and numerical details from individual accounting machine cards, or to print classifications and to accumulate and print totals, net totals, and accumulated net totals. To complete these business tasks, the machine was equipped with an automatic plugboard—similar in principle to a telephone switchboard—which basically meant that the desired arrangement of data could be obtained from the punched cards. It could list cards at the rate of 80 cards a minute. Introduced in 1934 and marketed by IBM until 1949, the successful production of 405s helped establish Endicott as IBM's "Main Street."

The IBM Endicott plant spawned a number of important products in addition to the System 360 and the 405 (e.g., IBM 602, IBM 650 RAMAC, IBM 709, IBM 1401, and some of the IBM System/370 processors). The post–World War II years were especially productive in the realm of electronic data processing, and the IBM 650 Magnetic Drum Data Processing Machine quickly became a symbol of "modern" computing. While the famous 700 series of IBM computers was being developed at the company's Poughkeepsie, New York, facility, IBMers in Endicott were making their own critical contribution to information technology history with the development of the IBM 650, which was referred to as "the workhorse of modern industry" and became the most popular computer in the 1950s.

Announced in 1953, the IBM 650 brought a new level of reliability and efficiency to the then adolescent field of electronic computing. One example of this advancement in computing reliability was that

Figure 3.2 IBM-Endicott Plant, circa 1960. IBM Corporate Archive.

whenever a random processing error occurred, the IBM 650 could automatically repeat portions of the processing by restarting the program at one of a number of breaking points and then go on to complete the processing if the error did not reoccur. That was a big improvement over the previous procedure requiring the user to manually direct the machine to repeat the process.

At the time the 650 was announced, IBM contended it would be "a vital factor in familiarizing business and industry with the stored program principles," and a vital force it became (IBM 2008a). The original market forecast for the 650 predicted that 50 machines would be sold or installed in 1955. By mid-1955, there already were more than 75 installed and operating, and the company expected to deliver "more than 700" additional 650s in the next few years. The year following this prediction, there were 300 machines installed and new 650s were coming off the production line at the rate of one every day. By the time manufacturing was completed in 1962, nearly 2,000 had been produced. At that time, no other electronic computer had been produced in such quantities (IBM 2008b).

IBM's critical role in all of these groundbreaking advances in modern computing would not have happened without Watson, Sr.'s absolute commitment to training and education. In May 1932, Watson, Sr., formally established an Education Department to manage IBM's many educational activities for employees and customers. A year later, in 1933, the IBM schoolhouse was built in Endicott to provide training and education for employees with a wide variety of expertise, including salesmen, engineers, tool-making apprentices, supervisors, and systems service women (Aswad and Meredith 2005). The IBM Schoolhouse still standing in Endicott is a reminder of Watson's commitment to education and IBM sustaining its leading role in the computing industry. Progressive educational opportunities became a major IBM priority, and Watson wanted this to be remembered each time employees and customers walked into the building, so he had his "Five Steps to Knowledge" philosophy engraved on the schoolhouse entry steps. Think, Observe, Discuss, Listen, and Read became the pillars of IBM epistemology.

IBM was one of the first American companies to hire and train women in engineering and technology work. I interviewed three women who worked for IBM for a short time during World War II, when many men were at war and women began to enter the labor force nationwide. One elderly woman I interviewed who worked at the plant for a short period in the late 1940s said that "IBM made sure its workers had a good wife. They really emphasized that. I suppose they felt this made better workers." Without a world war and a high demand for men, it is safe to assume that IBM would not otherwise have hired women for "skilled" technology work, as this was certainly a man's occupational territory, especially during the science and technology boom of the post–World War II era.

Many residents I interviewed brought to my attention that IBMers had "their own culture," that they had a tremendous amount of IBM pride and loyalty. This IBM culture manifested itself in many ways. For example, IBM established its own baseball team to bolster its corporate culture. The team grew out of ITR's earlier team, which was established in 1919, changing the name from the ITR "Ball Tossers" to the IBM "Tabulators." Additionally, IBM had a marching band that would perform at local events and company parades that would weave through downtown Endicott. With its marching band, company publications (e.g., THINK), company songs, strict dress code, and its own police force, one resident I interviewed contended that IBM's local corporate culture and power resembled a kind of corporate "fascism"[14] that has deeply informed local culture:

> IBM, and I don't know what you see when you look at the IBM workforce pictures, but it is a sea of white guys, you know. Certainly the employment of women changed things, but the dress code and everything is certainly telling of the culture of IBM here. I think it is very, very telling that all the photos from the 40s and 50s of all the employees crowding the streets and the IBM connections to the German fascist regime. There is certainly a feeling from those photos of a kind of fascism and paternalistic orientation and outlook that would eventually end up constituting the internal landscape of the people, you know.

These features of early IBM culture and history influenced the vision of Thomas J. Watson, Jr. Among other IBM principles in place, such as "Respect for the Individual," "Service to the Customer," "Excellence Must Be a Way of Life," "Managers Must Lead Effectively," "Obligations to Stakeholders," and "Fair Deal for the Supplier," Watson, Jr., added in 1960 that "IBM Should be a Good Corporate Citizen":

> We accept our responsibilities as a corporate citizen in community, national and world affairs; we serve our interests best when we serve the public interest. We believe that the immediate and long-term public interest is best served by a system of competing enterprises. Therefore, we believe we should compete vigorously, but in a spirit of fair play, with respect for our competitors, and with respect for the law. *In communities where IBM facilities are located, we do our utmost to help create an environment in which people want to work and live. We acknowledge our obligation as a business institution to help improve the quality of the society we are part of. We want to be in the forefront of those companies which are working to make our world a better place.* (IBM Corporate Archive, 2410MP03, p. 3; emphasis mine)

It is no mystery that IBM's positive (or lucrative) relationship to its stockholders built Big Loyalty in Big Blue, making IBM an early symbol of national corporate success, despite historic and ongoing IBM anti-trust cases filed by the U.S. Justice Department.[15] This fact was made clear at IBM's fortieth anniversary in 1954, when Watson's son, Tom Watson, Jr., delivered his speech to stockholders:

> I think the story can be summed up in six significant figures. In 1914 the gross income of this company was $4.1 millions of dollars. In 1953 it was $409.9 million. In 1914 the net before taxes was $489,000. In 1953 the net before taxes was $92.3 million. In 1914 cash dividends to stockholders were zero. In 1953 they were $12.7 million. Those six figures tell a story of a wonderful country, a wonderful business, and a most unusual and courageous leader. (Belden and Belden 1962:287)

IBM sold its personal computer (or PC) division, worth $1.75 billion in 2004, to the China-based Lenovo Group, a market currently dominated by Hewlett-Packard and Dell. Despite IBM's decision to shy away from PCs, it has created its own powerful position in the mainframe computer sector, a niche it began to develop in the early 1970s. Although mainframe R&D and production became a primary focus of IBM's Poughkeepsie, New York, plant in the early 1950s, the Endicott location began its focus on mainframes in the 1960s. It is said that today mainframes (or servers) continue their role in the era of e-business as the "back-office engines behind the world's financial markets and much of global commerce" (Lohr 2008:4). In fact, the majority of big businesses (Fortune 500s) worldwide rely on mainframe computers for their operations, and according to the Computer and Communications Industry Association (CCIA), mainframes support up to 80 percent of the world's electronic transactions, or computer-based transactions involving ATM sessions, airline bookings, tax filings, health records, and other essential services. IBM competes with other Fortune 500 IT firms that produce mainframes, like Microsoft and Oracle, but has, for the most part, dominated the mainframe computer business for the past 30 plus years. According to the CCIA, IBM controls nearly 50 percent of current mainframe industry profits.

Additionally, mainframes have been described as a "survivor technology" (Lohr 2008) that IBM has remained committed to, in order to serve the "secure" computing needs of critical sectors of contemporary capitalism:

The mainframe is the classic survivor technology, and it owes its longevity to sound business decisions. IBM overhauled the insides of the mainframe, using low-cost microprocessors as the computing engine. [IBM] invested and updated the mainframe software, so that banks, corporations, and government agencies could still rely on the mainframe as the rock-solid reliable and secure computer for vital transactions and data, while allowing it to take on new chores like running Web-based programs. (Lohr 2008:4)

By the 1980s, which is said to be the final heyday of IBM in Endicott, IBM employed in excess of 12,000 people at its Endicott facility. At one point IBM's employees in this part of New York State numbered close to 20,000. This figure, of course, doesn't include the hundreds of employees working for local businesses and the many smaller electronics firms that made different parts and components needed to bolster IBM's progress. According to the IBMers I interviewed, IBM's corporate downsizing started to impact local workers at the IBM facility in the mid- to late 1980s. Despite the early signs of these neoliberal economic trends, these same IBMers point out that it wasn't until the sale of IBM's facility in 2002 that IBM's "interests" really became a reality. As one IBMer put, "this was when the shit really hit the fan."

In 2002, IBM sold its property for $63 million[16] to a company called Huron Associates, and sold its 150-acre microelectronics division to Endicott Interconnect Technologies Corporation (EIT).[17] EIT, employing around 1,500 individuals, manufactures printed circuit boards (also known as PCBs), medical and advanced test equipment, and other electronic components for high-performance computing applications. While Endicott is the birthplace of IBM, it is now "home" to EIT.

Like IBM, EIT has developed and sustained its competitive edge through multiple business partnerships. These include strategic partnerships and contracts with SureScan (a competitive global company headquartered in Endicott that manufactures high-speed technologies for the homeland security marketplace), Cadence (a leading electronics manufacturer headquartered in California's Silicon Valley), Binghamton University, and Lockheed Martin.

EIT's partnership with Lockheed Martin was seeded by IBM's original Department of Defense contracts. The IBM Owego facility, just ten miles west of Endicott, opened in 1957 and was initially occupied by IBM employees who developed targeting and navigational systems for the Air Force's B-52 strategic bomber. Over time, IBM Owego grew to include programs such as defense systems for radar-evading stealth aircraft, computer hardware for the space shuttle and anti-surface warfare. With the downsizing of the Department of Defense (DOD) budget, the company set upon a course to diversify into non-DOD business areas. In 1991, aerospace products and programs comprised the majority of IBM Owego's business base. By 1997, nearly two-thirds of the company's business consisted of non-DOD programs, including postal systems, information systems, and international programs.

In 1994, the Loral Corporation, a high-tech company that concentrated on defense electronics, communications, and space and systems integration, acquired the IBM Owego facility. Lockheed Martin Corporation strategically combined with Loral's defense electronics and systems integration businesses in April 1996 and the Owego site became Lockheed Martin Federal Systems, Owego. In 1998, the site was converted into headquarters for Lockheed Martin Systems Integration–Owego, UK Integrated Systems, and Lockheed Martin Canada. Today, about 4,000 employees work at the Lockheed facility, a facility that, according to Lockheed Martin's Owego facility website, is known to be "a world leader in development and production of maritime helicopter systems, one of the largest suppliers of postal automation systems to the United States Postal Service, and a primary supplier of information technology solutions in the federal market."

Endicott's "Sleeping Giant": The Toxic Plume

While many residents choose to remember Endicott's industrial history for its business success and the village's status as the birthplace of IBM and George F's "industrial democracy," they also know that this history has been tainted by industrial pollution. Long-time residents remember the toxic smells of EJ's tanneries and the "mysterious odors," as one resident put it, that would seep out of IBM's smokestacks. Most residents point to IBM's reported chemical spill in 1979, while others more familiar and actively involved in the issue draw attention to other spills prior to the 1979 spill to explain IBM's tainted industrial history in Endicott.

The highly contested nature of quantifying the chemical spill(s) has led some to refer to the plume (or zone of contamination) as a "mythical beast." Here is the perspective of one former IBMer I interviewed:

> Between 1977 and 1978, as Grossman's book shows [Grossman 2006], the 4,100 gallons is hardly anywhere near what actually spilled. In February of 1978 alone there was a spill of 1.75 million gallons of industrial waste water, which IBM said had an unusual amount of methyl chloroform in it. IBM turned its head. I don't think people can even envision what 1.75 million gallons of water is. In one way or another either in the storm sewer vis-à-vis in the ground water. That's one spill. Spills of 80,000, spills of 200,000, spills of 120,000 in early 1978 that had all the heavy metals in it, the chromium, the hexavalent chromium, the copper, you know. Those amounts in 1977 and 1978 in a semiconductor operations expands four decades, you know. Who knows what's up. My sense about it is that what is underneath this community is egregious and so contestable at this point. So once again you have a situation where this concrete reality of this thing kind of in the underground and you can't see it so it is easy to skew the data and to create whatever you want with it. It is really a mythical beast and that is exactly what it is being relegated to. I mean there is every possibility that the levels underneath this community are in line with any other cancer situation.

Thanks to IBM and the New York State Department of Environmental Conservation, official records confirm that in 1979, IBM's Endicott facility spilled 4,100 gallons of the solvent TCA (1,1,1-trichloroethane, also known as methyl chloroform), a commonly used volatile organic compound (VOC). Described by some as Endicott's toxic "sleeping giant" (Grossman 2006:99)—the chosen words of congressman Maurice Hinchey—the news about the plume came from a comprehensive hydrogeologic report prepared by IBM indicating a larger than expected plume containing a toxic soup of industrial solvents, including trichloroethylene (TCE), tetrachloroethene (also known as PCE or perc), dichloroethane, dichloroethene, methylene chloride, vinyl chloride, freon 113. Years later, traces of benzene, toluene and xylene were also found in the local groundwater. Groundwater remediation and monitoring began immediately when IBM notified the New York State

Department of Environmental Conservation (NYSDEC) of the spill; this groundwater remediation effort continues today under the authority of a New York State Hazardous Waste Management Permit.

The spill contained a soup of chemicals, but the primary chemical of concern is TCE (trichloroethylene) and its breakdown products. TCE is a chlorinated solvent that was used during the 1960s and 1970s for metal degreasing and was commonly used in chipboard manufacturing processes. As a persistent volatile organic compound that is heavier than water and not very soluble, researchers find that when TCE enters groundwater sources it tends to persist for many years. Additionally, because of TCE's volatility, it can move from groundwater to soil and even vaporize, causing concern about TCE vapors and their impacts on indoor air quality. In fact, in response to this later concern and finding, TCE has been identified in at least 852 of the 1,416 sites proposed for inclusion on the U.S. EPA National Priorities List (Scott and Cogliano 2000).

According to NYSDEC officials, the degree of contamination is highest in the vicinity of the former IBM plant and diminishes with distance from the IBM plant site. Groundwater flow transports the contamination to off-site areas southwest of the plant, with lower levels extending as far as the Susquehanna River. The remediation effort that started after the 1979 spill resulted in the installation of pumping wells to capture the contaminated groundwater so it could be treated to remove the VOCs, including traces of TCE. A NYSDEC official told me in an interview that since the remediation effort began, about a million gallons of water are pumped and treated each day. Today, downtown Endicott's "plume" is dressed with monitoring wells and groundwater pumping stations. The NYSDEC is careful to communicate with the public the difficulty of cleaning up the contamination in its entirety, stating that "Although groundwater data indicate that this program has been effective, it typically takes many years, or even decades, to clean up groundwater." Like many other groundwater contamination risk communication strategies, the NYSDEC talks a lot about "natural attenuation," or the fact that pollutants slowly dilute over time. This is the "dilution is the solution" argument that many vapor intrusion activists, and other citizens engaged in similar environmental disaster zones (Button 2010), contest. For now, it is important to point out that most residents I interviewed believe that the TCE plume will never be completely cleaned up. "They have put millions of dollars into this cleanup and, you know what, at the end of the day, it may get smaller, but it won't go away. They [the NYSDEC] say this every time at the meetings, and I believe it," was the perspective of one resident who lives directly across from the former IBM plant and the point source of the first reported spills in the late 1970s.

While IBM has been evaluating ways to expedite groundwater source containment and VOC removal since at least 1980, the problem of "vapor intrusion" entered the contamination debate in Endicott when IBM sold its facility in 2002. The plethora of vapor mitigation systems visible on Endicott's landscape is the result of the state ordering IBM to mitigate homes in Endicott threatened by vapor intrusion, or the process by which volatile chemicals move from a subsurface source into the indoor air of overlying or adjacent buildings. The discovery of vapor contamination in Colorado at the Redfield site in the late 1990s led to a review of sites previously believed to have little potential for vapor intrusion. The IBM Endicott site was the first of these sites in New York at which it became clear that vapor intrusion from contamination was affecting residential homes and other buildings.

Indoor air sampling was initially performed at the IBM Endicott site in 2001 by IBM as part of state implementation of EPA's Resource Conservation and Recovery Act (RCRA) Environmental Indicators Initiative, and included concurrent sampling and analysis of groundwater, soil, and subsurface soil vapor. Subsequently, in the spring of 2002, both the NYSDEC and the New York State Department of Health (NYSDOH) required IBM to evaluate the potential for vapor intrusion into buildings over the roughly 300-acre plume of solvents linked to the former IBM campus.

An approach to evaluate the potential for vapor intrusion was developed by IBM in consultation with NYSDEC and NYSDOH and finalized in December 2002. The sampling plan for structures was designed to identify, during the 2002 and 2003 heating season, the extent of the area where mitigation systems would be offered. In order to accomplish that objective, IBM elected to sample approximately 20–25 percent of the structures above the 300-acre plume rather than every structure (NYSDEC 2008).

Stimulated by a growing understanding of the science of the vapor intrusion pathway, in 2002 the NYSDEC required IBM to investigate the potential for toxic vapors to migrate from groundwater through the soil into buildings above the plume. The results of the investigation indicated that vapor migration had resulted in detectable levels of contaminants in indoor air in structures, including off-site locations in Endicott and the town of Union to the west. TCE was and remains the primary contaminant of concern with respect to indoor air, and the state's TCE guideline for TCE in indoor air is 5.0 mcg/m^3 (micrograms per cubic meter). By 2004, IBM had identified approximately 480 "structures" at risk of vapor intrusion and began to offer owners vapor mitigation systems (VMSs) to intercept the IBM-related contaminant vapors. Since 2004, nearly 470 systems have been installed on 418 properties.

The indoor air concentration of TCE in structures offered VMSs ranging from 0.25 to 7.6 mcg/m^3 in indoor air and 260 to 4,400 mcg/m^3 under the sub-slab. Of the 144 samples taken, roughly 90 percent (130) had indoor air concentrations of less than 0.22 mcg/m^3. No structure that was not offered mitigation had an indoor air level above 0.67 mcg/m^3 or a sub-slab level above 250 (NYSDEC handout). TCE is one of 188 hazardous air pollutants (HAPs) listed under section 112(b) of the 1990 Clean Air Act Amendments, and currently the U.S. EPA has set a maximum contaminant level for drinking water of 0.005 mg/L (micrograms per liter), and the U.S. ATSDR has set minimal risk levels (MRL) at 2 ppm (parts per million) for acute inhalation and 0.1 ppm for intermediate inhalation (USEPA 2007).

Vapors can enter buildings in two different ways. In rare cases, vapor intrusion is the result of groundwater contamination which enters basements and releases volatile chemicals into the indoor air. In most cases, vapor intrusion is caused by contaminated vapors migrating through the soil directly into basements or foundation slabs. Although the NYSDEC historically has evaluated soil gas pathways, advances in analytical techniques and the knowledge gained from remedial sites in New York and other states has increased the NYSDEC's understanding of how vapor intrusion occurs. As stated on the NYSDEC website, the agency's understanding of the risk of the vapor intrusion exposure pathway has shifted over time:

Historically, we thought that vapor intrusion was only an issue where the source of the contaminants was very shallow and the magnitude of the contamination was very great. We now know that our previous assumptions about the mechanisms that could lead to exposure to vapor intrusion were not complete. The result is that additional work may be required to investigate or remediate sites that are in the operational or monitoring phase, or that have already been closed. Separate ranking systems have been developed to account for the two different sources of contaminated vapors. Because we now recognize the need to take a different sampling approach, when the [NYSDEC] evaluates a site for vapor intrusion, both sources can now be effectively considered.[18]

One major challenge for understanding vapor intrusion risk is that contaminated soil vapor is not the only possible source of volatile chemicals in indoor air. Volatile chemicals are found in many household products, such as paints, glues, aerosol sprays, new carpeting or furniture, refrigerants and recently dry-cleaned clothing that contain off-gas contaminants like perchloroethylene (also known as tetrachloroethylene or PERC). Volatile chemicals are also emitted by common commercial and industrial activities, and indoor air may also become affected through the infiltration of outdoor air containing volatile chemicals. As I observed during my fieldwork and as will be shown in chapter 6, many residents are critical of the fact that the VMSs don't have a carbon filter and wonder to what extent the TCE vapor "really" dissipates once it is released above the building or home, as the NYSDEC and vapor intrusion scientists tend to argue.

The NYSDEC and the NYSDOH have led the vapor intrusion risk communication effort in Endicott. Perusing their websites, I found that together they agreed that vapor intrusion was "among the top priorities of our agencies," that "the likelihood of vapor intrusion-related exposures varies from site to site," that "the number of sites at which vapor intrusion evaluations are warranted is quite large," and finally, that while Endicott presents a special case for vapor intrusion risk, "revisiting this issue concurrently at all volatile chemical sites where remedial or corrective actions have been implemented is not feasible, resulting in the need to prioritize these sites." I confronted a discourse of progressive, emerging science, a complex science in-the-making, with both agencies claiming that the "process for evaluating vapor intrusion exposure is evolving, as we learn more about this highly complex phenomenon. . . . Nonetheless, New York's efforts are leading the Nation in the development and implementation of strategies to address vapor intrusion." They add that "Through our efforts, our citizens will be better protected from the chronic health problems which can result from exposure to volatile chemicals in indoor air."

The state's vapor intrusion mitigation decisions vary from site to site. In fact, currently there is no national standard for vapor intrusion regulation, which was the impetus for the EPA's 2009 National Forum on Vapor Intrusion, which I discuss in more detail in chapter 6. But according to the NYSDEC website,

> it may appear that we [the NYSDEC] are applying our vapor intrusion policy and guidance inconsistently. In reality, however, decisions on how to address exposure to vapor intrusion will be made on a site-by-site basis, after a comprehensive review of individual subsurface vapor, indoor air and outdoor air results, and after the consideration of additional site-specific parameters, such as sampling results; sources of volatile chemicals; background levels; and applicable guidelines for volatile chemicals in the air. This is the most appropriate approach to ensure the protection of public health.

The NYSDEC and the NYSDOH view the use of vapor mitigation systems as a short-term solution to the vapor intrusion problem. Several residents I interviewed likened the installation of VMSs on homes and business in the plume to a "Band-Aid" that fails to fix the bigger problems, like having to live in a toxic plume with uncertain health risks and crippling property values. But, despite residents' critiques and concerns, the NYSDEC believes that addressing the source of the contamination and ensuring that steps are taken to remediate and monitor the soil and groundwater which provides a pathway for the migration of these chemicals, it "can provide effective long-term protection of the public health from vapor migration" (NYSDEC handout). Based on public meetings I attended in 2008–2009, the NYSDEC reports that sampling data indicate that there are sources of vapor contamination in addition to those associated with the IBM spill(s), including local dry cleaners and VOCs used at the former Endicott Forging Company.

The IBM-Endicott site was initially listed as a Class 2 Superfund site in 1984, then downgraded to a Class 4 site in 1986. Class 4 sites are determined to be properly closed, but continued management is required. Upon appeal by several parties, including New York State Congressman Maurice Hinchey and New York State Assemblyman Tom DiNapoli, the site was reclassified as a Class 2 in February 2004. The same day of this reclassification, IBM gave the village of Endicott a "gift" of more than $2 million.[19] According to the NYSDEC, the reclassification occurred as a result of "new information regarding groundwater contamination and soil vapor intrusion into structures in the area over the groundwater plume." This reclassification was significant because a Class 2 site is one where "hazardous waste constitutes a significant threat to the public health or environment," leading to an intensification of the government response and attention to the public health risk of the IBM spill and its remediation. The public health risk of TCE vapor is very ambiguous, although a 2012 study at the Endicott site has shown that residence in the TCE plume was associated with low birth weight and fetal growth restriction (Forand, Lewis-Michl, and Gomez 2012).

Issued in 2003, the NYSDOH set its air guideline for TCE at 5.0 mcg/ m^3 (micrograms per cubic meter). This guideline is highly contested by activists from communities impacted by TCE contamination, as will be explained in chapter 7, but what is important to point out here is that, according to the NYSDOH, the TCE data from the Endicott site and the Hopewell Junction site in Dutchess County helped determine the current TCE exposure guidelines for New York State.

In September 2004, the NYSDEC and IBM entered into a formal consent order[20] that requires IBM to investigate and remediate contamination in Endicott. This consent order requires IBM "to conduct a supplemental remedial investigation and focused feasibility study program for seven operable units that will identify and evaluate previously unknown or insufficiently evaluated potential sources of pollution at and in the vicinity of the site, and develop and implement appropriate cleanup measures." Additionally, several interim remedial measures (IRMs)[21] have been deployed to address known environmental contamination at and in the vicinity of the former IBM plant. The consent order gives NYSDEC the authority to require additional IRMs as appropriate and sets forth specific IRMs to be undertaken by IBM. One of these IRMs involves the use of a new remediation technique called "in situ thermal remediation," a process whereby the contaminated groundwater source is heated up, resulting in faster vaporization, and theoretically, faster remediation of toxic vapors. This is a remedial effort that is taking place across the street from the former Endicott Forging plant. I walked past this site many times during the course of my fieldwork. A fenced off science project that hopes to reveal a new cutting-edge remedial effort, the in situ thermal remediation project is expensive and reinforcing the finding that TCE is still very difficult to fully remediate.[22] One resident and activist I interviewed believes that this remedial experiment may help clean up the plume and new discussions of bioremediation options—injecting microbes that can metabolize volatile organic compounds like TCE into the contaminated groundwater source—are signs of things moving forward, but he is still "tired of living on top of a science project. . . . Science is great. I just don't think people should have to live on top of a science project."

IBM turns to techno-scientific expertise and virtues (Shapin 2008) to solve "complex" problems like vapor intrusion, as it does to address "societal problems": "IBM remains committed to solving societal problems through a range of programs that bring expertise and skills development where they are needed most. We approach these complex issues—from childhood cancer, to literacy, to entrepreneurial support— by looking at them systemically and engaging our global community of IBMers and our best technology and knowledge to reach scale."[23] As it turns out, the "societal problems" experienced and vocalized by Endicott residents are concretely linked to IBM deindustrialization and contamination and the lived community transformations resulting from these and other processes.

4

Living the Tangle of Risk, Deindustrialization,
and Community Transformation

Scholarship wrestling with the intersection between environment and health almost inevitably confronts tangles of
economy and flesh.
—Mitman, Murphy, and Sellers (2004:11)

Whether or not you end up with cancer, your quality of life
suffers and changes as long as you live here.
—Plume resident, proprietor, and retired IBMer

Endicott's IBM spill archive, located at the public repository in the
George F. Johnson Memorial Library in downtown Endicott, contains
four large shelves packed with fact sheets and technical reports, many
of which are developed by private companies contracted by IBM and
the NYSDEC to carry out the plethora of environmental and public
health analyses for Endicott's groundwater monitoring and remediation
project. These reports and fact sheets are filled with monitoring well test
data, soil gas sampling data, public health statistics, indoor air sampling
findings, analytical summaries, addendums, memorandums, and very
few public comment reports. What is not present among this inventory
of fact sheets, reports, and data, what is missing from this impressive
archive of "public" information, is in-depth narrative and survey data
reporting on the perspectives and experiences of plume residents coping with this microelectronic industry disaster. The next two chapters
explore this "other" data.

This chapter combines the analysis of qualitative and quantitative data to discuss the ways in which plume residents understand and make sense of the impacts of the IBM pollution they live with and in. What is the experience of residents living with/in the IBM plume? Amid IBM's boom and bust and over 30 years of remedial work, what do Endicott's plume residents feel certain about? What remains elusive? Do differences in age, gender, education, length of residence, and renter/owner status influence plume residents' perceptions of risk and their perceptions of the quality of life in a late industrial Endicott marked by economic corrosion and environmental destruction? These are the core questions to be explored in this chapter.

While most plume residents I interviewed tend to agree that ambiguity and uncertainty are ubiquitous—especially when it comes to "knowing" what is really going on with the plume, with vapor intrusion and the cleanup effort, and how TCE exposure is impacting peoples' health—there was far less confusion about the fact that IBM's pollution is a big problem in downtown Endicott. Most residents I interviewed and surveyed rightfully contend that TCE is a toxic substance to be concerned about, that stigma is something plume residents experience or "learn to live with"—as one resident put it—and that Endicott is one of many communities nationwide struggling with trenchant social, political, and economic symptoms of deindustrialization. The strongest theme to emerge in my discussions with plume and non-plume residents alike is that Endicott is on the wane.[1] Residents' reflections on and understandings of Endicott's tainted post-IBM landscape, in other words, bring to the fore the enduring interconnections between and inversions of economy, environment, health, and social life. Local discourses on the certain and uncertain impacts of the IBM plume invoked a rethinking of sorts, a renegotiation of Endicott as a desirable community and place to live.

In what follows, I first analyze narrative and survey data that draw attention to similarities and differences in local framings and perceptions of TCE contamination and vapor intrusion in general and perceptions of the health risks of TCE exposure in particular. We see how intersubjective experience informs plume residents' understandings of TCE risk, as well as how these risk understandings and perspectives expose certain contradictions and dynamics. Additionally, I show how residents' local health risk understandings are entangled in occupational health risk talk, particularly discourse on concerns about mortality trends for former IBMers who worked at the IBM Endicott plant, which is the focus of an ongoing National Institute for Occupational Safety and Health (NIOSH) study.

Another goal of the chapter is to analyze narrative and survey data on the intersecting themes of IBM deindustrialization, dystopia, and stigma. For many plume residents I spoke with, the TCE contamination is "just another sign"—as one resident put it—of Endicott falling apart and becoming yet another Rust Belt town to go bust after many years of industrial dynamism. Many discussed that they are witnessing an overall decline in the "quality of life" in Endicott, a community decline discourse that coupled deindustrialization and contamination processes, that focused on toxico-economic sorption. For a community that once thrived with families, businesses, and workers, it has now, for many, lost its lure and become a welfare community with growing class- and race-based tensions. Connected to these post-IBM transformations is a common emotive theme to surface in contaminated communities: stigmatization. While most plume residents I interviewed explained that living in the plume has amounted to a stigmatizing experience, it was not something that bothered everybody. For example, one resident I interviewed who was born in 1924 and has lived in Endicott for 82 years told me, "I don't feel any stigma. No. I mean what difference does it make? I mean, hey, show me a town anywhere where they can say 'We're green and we don't have a thing to worry about here.' This was the only explicit anti-stigma talk I confronted. Most plume residents I interviewed believed that being a "plumer," as one resident described herself, tends to mark you as being a member of a particular geographic space, a toxic zone with tainted property and ambiguous cancers. Most important, though, I found that residents' discourses on these experiences of deindustrialization, community change, and stigma were infused and not understood as mutually exclusive post-IBM effects.

The State and Science of TCE

To contextualize my analysis of community understandings of TCE risk, I wish to begin with a brief look at how the toxicology of TCE[2]— the primary chemical of concern in the plume—is being framed beyond the Endicott community, at the state and federal levels. Federal and state public health and environmental agencies all recognize trichloroethylene or TCE as a hazardous chemical, especially since it is one of the most common toxic substances to be found at Superfund sites nationwide. According to the EPA, it is one of the chemicals most often released into the environment (U.S. EPA 2001). As a nonflammable chemical solvent most commonly used for degreasing of metals, TCE was a popular industrial solvent for many years and heavily used in microelectronic production. In fact, TCE was the targeted solvent of electronics workers in Silicon Valley in the late 1970s, when they organized to form the Santa Clara Center for Occupational Safety and Health (SCCOSH) to ban TCE from the workplace.[3] TCE is highly volatile, so when released into the environment it most often finds its way into the air. When TCE is released into groundwater, it can persist for many years (or decades) and lead to slow vapor intrusion scenarios, as is the case in contaminated communities like Endicott. When people are exposed to TCE, it is readily absorbed by all exposure routes and is widely distributed throughout the body.

Both the U.S. Agency for Toxic Substances and Disease Registry (ATSDR) and the EPA recognize the health risks of TCE exposure to humans. Knowledge of the potential health risks of trichloroethylene (TCE) in the environment has resulted in it being listed as a "chemical of concern" across numerous environmental and public health programs. TCE is one of many hazardous air pollutants listed under the Clean Air Act. It is also a toxic pollutant under the Clean Water Act, a contaminant under the Safe Drinking Water Act, a hazardous waste under the Resource Conservation and Recovery Act (RCRA), and a hazardous substance under the Comprehensive Environmental Response, Compensation, and Liability Act (CERCLA or Superfund). It is a toxic chemical with reporting requirements under the Emergency Planning and Community Right-to-Know Act and the Toxic Substances Control Act. TCE releases, in other words, must be reported to the Toxics Release Inventory. Community-right-to-know programs also include TCE on their toxic substance lists. Cancers that have been linked with TCE exposure include cancers of the liver and bile ducts, kidney, and esophagus, as well as non-Hodgkin's lymphomas (Chiu et al. 2013; Mandel et al. 2006; Wartenberg et al. 2000); one epidemiological study at the Endicott site has associated residential TCE exposure with low birth weight and fetal growth restriction (Forand et al. 2012). But, like most toxins, TCE's cause-effect relationship is caught up in social debates over science and expertise. Furthermore, TCE risk is represented and communicated differently by different parties engaged in TCE politics.

In 2005, the "TCE Blog" was developed by a concerned resident from Chesire, Connecticut, a community that has been struggling with TCE contamination since the late 1970s. The goal of the TCE Blog is "to provide a consolidated online resource for individuals and communities to learn more about TCE, TCE contamination, and the apparent potential obvious relationship between TCE exposure and adverse health effects."[4] The words "apparent" and "potential" were crossed out on the blog's website in 2006 after information was released from the National Academy of Sciences' review of the EPA's controversial 2001 draft risk assessment of TCE, which showed that TCE exposure was actually linked to more health effects than originally reported by the EPA. When I attended an expert workshop on vapor intrusion in Portland, Oregon in 2009, the discussion of TCE vapor intrusion (VI) was dominated by the process by which volatile organics migrate and enter homes. The focus was not on the environmental epidemiology of TCE vapor intrusion. This is an area of knowledge that is poorly understood, but this does not mean that residents of communities like Endicott wait on epidemiological evidence to decide their own level of security. The fact is that existing circulation of information on TCE risk informs how residents and activists "think" about and reflect on TCE risk. In some ways it motivates a certain "willingness to endure a condition of mental unrest and disturbance" (Dewey 2005 [1910]:12), no matter the degree or number of environmental epidemiological studies and expert assurances.

The TCE risk debate continues to unfold in both science and toxics advocacy circles. Members of communities impacted by TCE contamination have recently organized to sign a petition to augment the EPA's efforts to create a more credible TCE health risk assessment. Lenny Siegel, a leading TCE and vapor intrusion activists with the Center for Public Environmental Oversight, helped craft this petition, sent to Lisa Jackson, the EPA's current administrator, on January 15, 2010:

Dear Administrator Jackson:

We are people from communities impacted by trichloroethylene (TCE) pollution, or we represent such people. We are pleased that EPA has once again put forward a Toxicological Review, in support of the

Integrated Risk Information System (IRIS), for this compound. We urge you to continue to resist pressure to defer action by sending the question of TCE toxicity to the National Academy of Sciences (NAS) for yet another review.

As you know, the NAS published its last report on TCE in July 2006. After a more than 18-month effort the NAS found, "The evidence on cancer and other health risks from TCE exposure has strengthened since 2001." Furthermore, it recommended, "enough information exists for the U.S. Environmental Protection Agency to complete a credible human health risk assessment now." That's exactly what EPA should do.

We recognize there is a need for additional research on the health effects of TCE exposure, but as the Government Accountability Office has repeatedly pointed out, delays "can result in substantial harm to human health, safety, and the environment." That is, we don't want our families and our communities to be "guinea pigs" in a permanent research experiment. Innumerable people are being exposed to TCE levels, in the air we breathe and water we drink, that meets current, outdated standards but which would be found unacceptable under EPA's draft Review.

Moreover, in May 2009 your office issued a new "Process for Development of Integrated Risk Information System Health Assessments," streamlining what had become a cumbersome, unproductive process. The TCE Review is one of the first tests of this process. Failure to complete the TCE Review in a timely fashion would undermine the entire IRIS system, setting a precedent for the continuing delay in the development and improvement of science-based exposure standards.

Thus, we urge EPA to consider carefully the comments its receives on the draft TCE Review and to act quickly to finalize new IRIS values for TCE based upon today's best science.

As stated earlier, the struggle for Endicott's plume residents goes "beyond the health exposure" concern, beyond the probabilistic science of risk, which is itself vexed by uncertainty.[5] On the other hand, that is not to say that residents living in and with Endicott's IBM TCE plume are not concerned about the health risks of TCE exposure or that they don't attempt to make sense of these risks based on their own life experience. Amid grassroots efforts to create TCE risk standards based on "today's best science," on "sound science," residents also make sense of TCE risk and the uncertainty of TCE risk in their own terms and based on their own local and situated knowledge (Geertz 1983; Haraway 1988).

Concrete Experiences and Understandings of TCE Risk

A central focus of my research in the IBM-Endicott plume was to better understand how Endicott residents living in the plume thought and talked about the health risks of TCE contamination. To flesh out this understanding, I collected in-depth narratives of plume residents and surveyed households in the plume to get a sense for how they understood the health risks of TCE exposure. One of my primary goals for conducting a quantitative survey of plume residents, aside from generating some basic demographic information on characteristics of the plume population, was to explore whether residents living in Endicott's TCE plume had a high sense of environmental health risk. I was curious to know how residents ranked the connection between IBM pollution and public health risk in Endicott. I found that the majority (64.6%, or 51 out of 79 respondents) of those surveyed (n=82) "strongly agreed" with this statement. The survey then asked residents whether or not TCE vapors in particular had impacted local public health, and again the majority (65.4%, or 51 out of 78 respondents) "strongly agreed." For survey respondents who marked "strongly agree" to this second scaled question ("TCE vapors have impacted local public health"), only seven respondents marked "yes" when asked the question "Have you had any personal health problems that you believe are related to TCE or other toxic substances resulting from the IBM spill?" The "personal" health problems reported by these seven respondents were "I am having female problems," "My health has deteriorated substantially since living in this home. I have fatigue and nausea," "I get headaches, sore throat, breathing problems, and also emotional suffering. It is stressful not knowing if I or my family will be another victim of this IBM negligence," "I have neuropathy," "I am having a lot of problems with skin irritation and upper respiratory issues," "I have asthma in the Spring when everything in the air is blowing around," and "I have cancer." The survey also asked plume residents to list what they thought were the health effects of TCE exposure, to see how they compare to the EPA's most recent environmental epidemiological findings.

Table 4.1 Health Effects of TCE Exposure: Plume Residents vs. EPA Experts

Plume Resident Understanding	EPA Understanding
Cancers	
Respiratory problems	Kidney cancer
Birth defects	Liver cancer
Heart problems	Lympho-hematopoietic cancer
Neuropathy	Cervical cancer
Nervousness	Prostate cancer
Skin problems	

Only 23 of those surveyed (n=82), responded to this question, and most either said "don't know," "not sure," or in one case, "I have no idea, but with all the money IBM is spending to clean it up, it can't be good." Some respondents, though, did list what they felt were the health effects of TCE exposure (see table 4.1). While it is true that these TCE risk perceptions or the so-called "social effect of risk definitions . . . [are] not dependent on their scientific validity" (Beck 1992:32), it is worth remembering that TCE health risk knowledge is still in the making, as are residents' senses and understandings of TCE risk. Both ways of knowing matter, both emergent realities matter to risk decision-making and systematic knowledge formation (Wynne 1989; Jasanoff 1992). To get real with the actual TCE risk situation, in this sense, is to openly recognize that exactness isn't necessarily or ever the dominant ingredient of lay or expert motives and rationalizations.

On the other hand, and this is where environmental health politics really figure into TCE contamination debates, many environmental epidemiologists would consider the locally defined environmental illnesses to be examples of anecdotal evidence.[6] Plume resident discourses on TCE risk might in fact be a case of people "confusing cause and effect," people engaging in "the real corruption of reason," as Nietzsche put it long ago in his 1888 *Twilight of the Idols*. But, this book is not interested in "scientizing" (Wilce 2008; see also Hacking 1999:105) these environmental health claims and narratives; that motive has proven to befuddle even the most cultured environmental public health experts (Little 2009). What is the focus here is showing how TCE risk is socially experienced, how risk understanding is socially constituted, and how TCE risk is made a matter of concern for plume residents. This approach seems to speak more directly to the nature of risk *experience*, or the actual lived experience of TCE risk that plume residents navigate day to day. Following Lupton (1999), I am most interested in documenting plume residents' understandings of risk and how these understandings are "constantly constructed and negotiated as part of [a] network of social interaction and the formation of meaning . . . [T]he primary focus is on examining how concepts of risk are part of world views" (Lupton 1999:29–30). That said, TCE risk understanding is primarily of the "interactive kind" (Hacking 1999:105), of a micro-ecological kind. TCE is volatile and migratory; it is not a stationary toxin. The same can be said for TCE risk understanding. When studied ethnographically, risk is interactively understood and socially processed and judged.

While only seven survey respondents listed "personal" health issues that they believe to be related to TCE exposure, I found it much more common for the plume residents to "know" neighbors or have family members with certain illnesses that leave them asking etiological questions, thus informing their "public" health risk perception. Local risk perception seems to be most conditioned by local social processes— most notably intersubjective experience—rather than residents' beliefs about their own personal health problems. Knowledge of the health problems of others was a strong focus of survey respondents' perceptions of risk. Here are some examples: "My husband died two years ago with bone cancer. We had a bicycle shop in the basement for 15 years prior to this. I believe this TCE contributed to his cancer. He worked in the basement everyday"; "My step-sons who grew up here were both born prematurely"; or "My daughter all of a sudden has colds and stiffness. When she is not here she is fine." One survey respondent added:

> We get the headaches. We get a lot of headaches from living here. We don't know what it is caused by. It could be the natural gas. It could be a series of things. My doctor said this is not the healthiest area to live in and that they get a lot of patients with complaints. They get a lot of complaints from other residents living here. Some of the residents that I have talked to who have moved out of Endicott say they don't get the headaches anymore. They say that since they left they haven't had the headaches. So, there is a problem here.

While these brief comments of survey respondents can create a snap-shot of how people feel about the health risks of TCE exposure, I found that the narratives that emerged from my in-depth ethnographic inter-views with plume residents illustrate in more detail the "social" risk perception of plume residents. I found that my survey questions based on scaled questions lacked the depth provided by risk narratives, or more appropriately, risk testimonies.[7] For example, in my interview with Shannon, a plume resident and nurse, she immediately began to talk about her husband's health problems when the conversation turned to the topic of TCE risk:

> My primary concern is health. I mean my husband has been through cancer twice. I just really question, number one, are they truly getting it out, and what the long-term health effects of it are. You know. We have been here 21 years. I mean thank God I am ok, but my husband has been through it twice. I mean I had 8 children living in this house, but I really question down the road what is this leading to? If anything, I mean are they just grasping at straws. I don't know. I'm obviously not knowledge-able enough in the overall picture, but when the health department stood there and said to me "Well, we are only checking certain kinds of cancer." That to me is a crock, because cancer is cancer. It is the abnor-mal reproduction of cells. What difference does it make in who, when, where, or how. You know.

Shannon's understanding of environmental health risk is conditioned by health problems that are, as she puts it, "close to home":

> When it is so close to home, it really makes you think. I took care of my one dear friend, she was 85 years old. She was old. She was diagnosed in October of 2006 with the same cancer my husband had and she was dead in February. They said it was a rare kind of lymphoma that she had. But, her house was one of the ones tested and what they found on the basis of her house was high levels. I don't know the levels, but they were high.
>
> I went to one of the town meetings and they said it was the law that every doctor has to report all cases of cancer. I said "Well then doesn't it make sense when you've got 12 cases of cancer in 16 homes." I mean the young man next door was in his early thirties when he got it. His dad,

maybe fifteen years later, was diagnosed on a Monday with a stomach cancer and by Friday he was dead. That is why I question the idea that only the elderly are affected. I don't think a thirty-year-old should go like that. One the house on the corner, the green house, his brother was 21 years old and died of brain cancer and a year later their mom died with recurring breast cancer. We had a girl that lived across the street that died last May. She fought with breast cancer that went all through her. She left behind a four year old. This is just chance? I don't think so. I mean yes, the old lady I took care of, yes, she was 85 years old. I don't think age should be an issue. I mean that's my concern. I mean are we getting the whole story?

Shannon then went on to contest the fuzzy position of public health experts. She found it frustrating to have experts say that plume residents are living "within a safe range" even while reports (Chiu et al. 2013) are "finding" significant relations between TCE exposure and her husband's cancer:

The reports said "you don't have to worry about these and this is what you have to worry about. You're within a safe range." I mean how do they really know . . . Initially they first told me, when I questioned it and I brought out the cancer and I brought out my husband, they said it publically at the meeting: "There was no connection between these chemicals and non-Hodgkins lymphoma." Not two weeks later there was a report put out in the paper and they were finding a significant relation between this and non-Hodgkins lymphoma.[8] You know, I mean it's kinda frustrating. It's frustrating to be laughed at. You know, I don't claim to know everything, but I'm not stupid. I think that is what I resent more than anything else. It is that what we thought were legitimate concerns have just been ignored.

Shannon experiences the same "double bind" situations that many residents of contaminated communities find themselves caught in. She is between messages, dealing with "a persistent mismatch between explanation and everyday life" (Fortun 2001:13), between being told the environmental health risks are slim and knowing certain things are present and true (Checker 2007). Regardless of the state of the science of TCE and the continued emergence of new research findings, there seems to be an enduring sense of helplessness among plume residents. Most plume residents—with the exception of many renters, as will be discussed later—understand that they inhabit both a fragile environment saturated with ambiguity and a social environment that can lead to frustrating interactions with government officials, like public health experts with the Department of Health. In Shannon's words:

> I guess when you have gone through cancer twice, you just feel like you are on egg shells. Now, is it definitely a result of it, we don't know. But, it sure does seem, you know, a higher percentage in this area and not just one kind of cancer. I think the very first time I responded to a questionnaire and I think it was in the Press, and it was before this area was announced. They were looking for people to call in. I believe it was over in Endwell were they had several cases of pediatric cancers, like brain tumors and leukemia, and just terrible stuff. It was in a very high percentage of children. I called the department of health and I said "You are looking for some input on this and have you checked the water?" They thanked me and said "We appreciate your concern, but right now we are checking on pediatric cancers." I said "I am not trying to depreciate the seriousness of children having it, but when you have got another part of town, you know, so close and with so many varied kinds." I said "What difference does it make whether its children or its adults. Cancer is cancer. It kills." They again thanked me for my concern and said "We'll make a note of it." It was probably about a year or two later that we started hearing about the spill in this area.

In my conversations with plume residents in Endicott, it became clear that not all plume residents were equally concerned about the health risks. For example, Lenny, Shannon's husband, is sure that he will never really know how he got cancer twice, but he does not rule out the "possibility" the plume has caused some local health problems:

> I had cancer twice. Non-Hodgkins Llmphoma. Twice. The first time was in 1993 and the second time was in 2002. Nobody knows the cause of cancer. Could it have something to do with the toxins that are here? Sure, it could, but no one has proof. It is just really unproven. I mean you can't really, I mean, they know through lab tests and stuff like that TCE can

cause cancer, but is that what my problem is? Well, that is something you'll never know . . . We just don't know what the health effects are. There is just no conclusive proof that this is the culprit. The only thing they say they know about TCE is from studying its effects on rats. Well, what does that mean? It is certainly a possibility.

Lenny's good friend Robert is, like Lenny, not sure why Lenny has had two episodes of cancer. They both accept the fact that many variables contribute to ill health. But, Robert is confused about why the ATSDR-DOH health statistics' review study of plume residents did not look at lymphatic cancers like non-Hodgkins lymphoma:

When you are dealing with people who are living in the plume, there are so many other variables. What are the health risks that the particular individual has? Larry is a case in point. He has had two bouts with cancer. The study has been done and the statistics have been run for the houses just south of the IBM complex, looking at the incidence of cancer. It shows some signs of slightly elevated cancers, but they did not look at lymphatic cancer at all. That is an issue that I don't understand. Why did they not look at the cancers of the lymph nodes? That was what Larry had, so I just don't know. My wife is more medically in tune. There must be a reason why it was left out, but I don't understand what it is.

Ambiguity can be a common source of struggle for residents of contaminated communities (Auyero and Swistun 2009; Edelstein 2004). Health impact is "hard to prove," leaving many plume residents—even those with family members who have specific cancers that are commonly the focus of environmental illness debates—feeling perplexed and stuck with concerns, questions, and the problem of not knowing:

When it comes to health, that stuff is hard to prove. That is really hard to prove. I mean personal activity plays a role. My brother got leukemia and they couldn't prove that it was the water. He died in 1985 when he was just 39 and there were 18 cases of cancer on that street. The people that lived in the house before him died of cancer. Their dog even died of cancer. There were different forms of cancer on that street. . . I just don't know what this stuff is doing to our health. That is the main concern.

What about our health? You don't know what it is going to do to your health.

In this way, not knowing or sensing the risks of TCE exposure, or having a firm belief that one is safe from TCE harm, as we see here, doesn't mean that plume residents don't witness other residents' struggles with illness which may or may not be caused by TCE exposure. This theme emerged in other interviews with plume residents. Ann, who worked at IBM for 18 years as a consignment analyst and now makes a living as a massage therapist, explained to me that even though she has witnessed her neighbor struggle with cancer, she still feels safe living in the plume. She saw the crippling effect cancer had on her neighbor and she explained that she gets excited now whenever she sees him walk the dog around the block, because for months, she says, "he was totally bedridden." Despite these observations and her belief that the plume is a problem and that "IBM has raped and ravaged this community," she is still unsure that the TCE is causing health problems or that it should be considered a valid cause for fear. As she cautiously put it, "With plume type things, you don't know what effects are here, in the ground, where it is. I am so afraid because I don't wanna like disregard or pooh-pooh someone else's fears and thoughts, you know." She then added, "But, you know, I mean I had a good reason to buy a $400 air purifier. You know, my thought was if I have a good air purifier, no matter what, if there is a problem, I'm protected. But others have not been so lucky."

Ann is not the only plume resident to feel unharmed, despite knowing "others" whose health may have been affected by the TCE plume. These two elderly residents are more concerned about the "risk" of property devaluation, but that does not mean that they do not think about the possible future health problems that may be the result of living in what they call "the heart of the plume" since 1970. Again, knowing of other plume residents struggling with cancer informs their sense and understanding of risk:

> ROGER: We haven't had any health problems and we have three children and they have not experienced any health problems either, so I guess our main concern is the property value . . . I guess I also don't worry because I don't know enough to worry about the health stuff.

HELEN: Of course the health issues may come up later, but right now, nothing.

ROGER: Not like the folks a couple blocks over, that's for sure.

HELEN: If you were talking about families on McKinley Ave and Monroe Street, than that is health problems. There is a lot of cancer there. There are something like five cancers in seven homes or something.

As Beck has argued, risks associated with chemical contamination "completely escape human powers of direct perception. The focus is . . . on hazards which are neither visible nor perceptible to victims" (Beck 1992:27). TCE risk might be imperceptible, but cancer, according to many of the plume residents, isn't. It is visible to many because many residents in the neighborhood suffer from cancer. But, plume residents are not alone. In an interview with a lead NYSDEC official working on the IBM-Endicott site, I was told that health problems are ubiquitous and impact everybody, and not just residents living in the plume. He spoke very matter-of-factly and with a quantitative tone, a statistical sensibility:

The fact of it is that something like one out of two or one out of three people get cancer. But if you had cancer, you know. My wife had cancer. Everybody has somebody in their family that has had cancer or that had some sort of medical problems. When you superimpose that relationship onto a community where there is contamination, the net result is that folks look at that contamination as the causal agent. You know, it is possible that it is in some cases, but there is probably a disproportionate kind of concern over the impact of the contaminants relative to what's really there.

What is really there, in the 300-plus-acre plume? What is the "net result" of local environmental health risk? Neither expert nor lay will likely find satisfactory answers to these questions. Scientists and residents alike know that cancer most likely has multiple "causal agents." The belief that "the possible is richer than the real" (Prigogine 1997) seems to fit well with the majority of plume residents' understandings and framings of the health risks of TCE exposure. Especially on the topic of cancer, possibility talk was far more common than certainty talk. Indeed, it is possible that TCE risk has "*different* meanings for *different* people" (Beck 1992:26; emphasis in original) and similar meanings for similar people. Differences in gender and renter/owner status, for example, seem to condition local knowledge of environmental contamination and residents' understandings of the TCE vapor intrusion process.

Experimenting with Risk Survey Data

While residents' personal narratives really anchor my anthropological interests, I did experiment with quantification during my fieldwork in the IBM-Endicott plume. Trusting that other environmental anthropology risk perception studies that employ quantitative methods (Tilt 2006) had something meaningful to say, I carried out a quantitative household survey in the plume to see if there was any statistical significance when comparing residents' risk perceptions across groups in the community (e.g., gender, owner/renter status, education, and length of residence). In this section of the survey I asked questions in the form of statements designed to measure each respondent's level of agreement. These statements (e.g., In Endicott, the IBM pollution affects local public health; or, The trichloroethylene (TCE) vapors found in Endicott impact local public health) were also combined with questions including, but not limited to: What do you think are the health effects of TCE exposure, or Have you had any personal health problems that you believe are related to TCE or other toxic substances resulting from the IBM-Endicott spill?

To make comparisons for groups with only two categories in the independent variable (e.g., gender), I ran a Mann-Whitney test using the Statistical Package for the Social Sciences software (SPSS). The Mann-Whitney test, also called the rank sum test, is a nonparametric test that compares two unpaired groups (e.g., male/female, renter/owner, etc.). When SPSS runs the Mann-Whitney, it ranks all the values from low to high. The low in this case is a 1 for "Strongly agree" and the high is "Strongly disagree." When it ranks these values, the test pays no attention to which group (e.g., male/female, renter/owner, etc.) each value belongs. If two values are the same, then they both get the average of the two ranks for which they match. The smallest number gets a rank of 1. The largest number gets a rank based on the total number of values represented in the two groups. The Mann-Whitney test sums the ranks in each group and reports the two sums, or two mean ranks. If the sums of the mean ranks are very different, the P value will be small, or preferably, close to <0.05. The result of the P value is the critical number for the Mann-Whitney test. The P value determines whether or not the populations or groups being compared have the same median, and the chance that random sampling will result in a sum of ranks as far apart (or more so) as observed in the actual experiment. In other words, if the P value is small and close to <0.05, I can comfortably assume that the results are not coincidence and that the populations do in fact have different mean ranks. I also ran a Kruskal-Wallis test on other independent variables (e.g., length of residence, age, and race) to see which variables explained differences in risk, but the ones I will discuss here are the only ones that showed statistical significance.

After running the Mann-Whitney test, I found only two scaled questions that showed statistically different perceptions of risk between males and females and homeowners and renters. When asked to rank, based on a scale from 1 to 5, with 1 being "Strongly agree" and 5 being "Strongly disagree," the survey question, "Environmental contamination is a problem in Endicott," I found statistical significance (p=.006) to be gender based. Women had a mean rank of 35.50, and men a mean rank of 46.64, which means that women showed higher perceptions of risk than men. What these results reveal is that men and women differ in their perception and that this difference is not coincidental. These results, in many ways, coincide with the literature accounting for gendered differences in risk perception (Cutter, Tiefenbacher, and Solecki 1992; Davidson and Freudenburg 1996; Flynn, Slovic, and Mertz 1994; Gustafson 1998; Slovic 1992).

According to Gustafson (1998:805), "[w]hen gender differences appear in risk research, they are generally expressed in purely quantitative terms and rarely related to relevant gender theory. This gives rise to serious problems regarding what gender differences are found, and how they are explained." Gustafson (1998) adds that "[q]ualitative differences, involving the perception of different risks and different meanings of risks, have been neglected or reduced to quantitative differences." While I don't wish to be included in this later camp, I do not think this book revealed enough ethnographic data on gendered differences to "speak to" the survey findings. At this juncture, I am comfortable siding with Davidson and Freudenburg (1996), who highlight how social roles and everyday activities can contribute to gendered differences in environmental risk perception. They conclude that since women generally tend to play the role of nurturer and primary care provider, women are often more concerned about health and safety. I would argue that while this could surely be a social risk perception fact, risk perception is more likely simultaneously informed by other social and intersubjective factors, such as power, socioeconomic status, and trust (Flynn et al. 1994). Again, my survey data showed a significant gendered difference, but my narrative data illustrated that both men and women had TCE risk concerns and felt environmental contamination was a "problem."

My survey results also showed significant differences between renters and homeowners when they were asked to rank the degree to which they understood the TCE vapor intrusion process.[9] Renters didn't "understand" the basic VI process, while homeowners did. I found this finding especially interesting, because while there is a New York State Tenant Notification Law requiring landlords to inform renters of environmental contamination problems impacting properties in communities like Endicott, renters are not necessarily being informed. Owners clearly have a better understanding of the TCE vapor intrusion process, and this tenant notification bill (A.10952-B/S.8634), which was signed into law by New York State Governor David Patterson in September 2008 and created by state assemblywoman Donna Lupardo, does not seem to be "informing" renters. After the bill was signed into law, Lupardo explained to reporters, t "It's only fair that we protect individuals and families who rent just like homeowners are protected when they purchase their homes."

While my survey results showed no statistically significant differences in health risk perception when comparing IBMers and non-IBMers, my interviews with former IBMers did reveal concerns about so-called double exposure scenarios. As one former IBMer put it, "there were some residents who were double whammied. They were exposed to it at work and they were exposed to it at home." Other residents felt that workers experienced greater risk: "With the plume, I mean, there are so many questions to ask. It is one thing for the people living in the community, but if you are talking about . . . workers who were exposed eight hours a day, five days a week for however number of years, it's a different issue. Let's put it this way. There is a good reason why NIOSH is doing their study. Now, what will the outcome of that study be? Who is to say?"

Uncertainty and "imperceptibility" (Murphy 2004) were risk themes that I expected to confront at the outset of my return to Endicott for fieldwork in 2008. Another theme of the research that surfaced with equal force was that of citizen struggles with deindustrialization, a topic compounded by the fact that this research was carried out during 2008, in the fog of the so-called "great financial crisis" (Foster and Magdoff 2009) which only exacerbated many residents' sense of social and economic insecurity and distress. Plume residents may struggle to understand the environmental health risks of TCE exposure and unusual rates of cancer in the community, but they were far less perplexed and confused when it came to talking about the reality of Rust Belt deindustrialization in general and IBM neoliberal creative destruction in particular. This experience of "insecurity" is now a national trend that is far from receding. As Besteman and Gusterson (2010:6) put it,

Americans sense that something is terribly amiss, even if the full picture is not entirely clear. Downsizing has devastated communities across America, replacing stable unionized jobs with low-wage insecure service sector jobs. The 2008 real estate crash and stock market collapse wiped out home equity and retirement savings while increasing unemployment, thus exerting further downward pressure on wages. But Americans have lost more than wages and equity during these last few decades; a more equitable division of risk among all income categories has been lost as well. In the decades after World War II, Americans could expect support from their employers and their government for managing the risks of illness, unemployment, and retirement, but since the 1980s, responsibility for managing these risks has been shifted onto the shoulders of American families. This "Great Risk Shift," says Yale political scientist Jacob Hacker, is "the defining economic transformation of our times."[10] The number of employees who receive healthcare benefits, defined-benefit retirement pensions, and benefits from unemployment insurance has plummeted over the past few decades. Whereas in most other industrialized countries the government provides benefit security, American families are left on their own as employers have withdrawn coverage. Americans are feeling insecure because they are.

Coping with IBM Exodus

During the first week of my return to Endicott in 2008, I was told by one resident and activist that "People are people. A lot of people just wanna talk to their neighbors about other things, like the kids, the weather, the new car. Nobody wants to talk about the shadow hovering over this community, you know." I agreed with this then, and to some extent, I still think this is true. Dwelling on Endicott's tainted IBM heritage is depressing. The "shadow" is likely not the theme of everyday dinner conversations. But, that is not to say that it is a topic devoid of local importance, or a discourse lacking local circulation. When I started to talk to plume residents about their situation and inquired about what it was like living in Endicott and what changes they had witnessed in the community, residents spoke at length about the local rupture invoked by the "shadow" of IBM deindustrialization and the inversion of Endicott's socioeconomic situation. All of the plume residents I spoke with and who responded to the plume community survey seemed to agree that since at least the early 1990s—when IBM was under the contentious and indecisive leadership of John Akers[11]—Endicott has been gradually experiencing a "community inversion" (Edelstein 2004:102) marked by widespread social and economic downturn. My neighbor Bill put it this way: "The quality of life here is depressed. You will find this in your studies. The village of Endicott has been in decline here because once the IBM gravy train folded, everything got depressed." Another resident, when asked "what is your vision of where the community is going?" responded simply with "Down."

My survey questionnaire included several questions that aimed to get at the theme of deindustrialization and Endicott's' "downward" swing. For many residents, the mixture of the IBM bust and the IBM contamination have made Endicott a less "desirable" place to live, with most plume residents surveyed (60.8%, or 48 respondents out of 79) blaming the IBM pollution for impacting the quality of life in Endicott. When asked if local industrial pollution has made Endicott a less desirable place to live, the majority of plume residents surveyed marked that they "strongly agree" (47.5%, or 38 respondents out of 80) or "somewhat agree" (42.5%, or 34 respondents out of 80).

Residents were also asked to rank the degree to which they felt the decline of local industry had made Endicott a less desirable place to live, to which over half (53.8%, or 42 respondents out of 78) marked "strongly agree." One resident's narrative actually points to the fact that the declining appeal of Endicott "had nothing to do with the spill." As this former IBM software engineer put it,

It is just an economy thing. When we first moved here it was great. IBM was going great. They were like smoking guns. Everybody was busy and a lot of things were going on. It was a great community to be in. But, when IBM slowly started going down and down and downsizing and downsizing, of course, naturally, that hit the economy of the area. Washington Avenue used to be a booming street for business. Now you look at Washington Avenue and you say "Ugh." I mean it is nothing like it was when we first moved here. There are still businesses there, but not like it was. Of course with IBM out of here, the job market here has just gone down the tubes. The jobs just went away. I mean there is still some IBM software stuff here, but it has downsized considerably and the only manufacturing still left here is EIT. What happened when IBM started to move out was that both software and hardware people were either relocating, retiring, or taking early buyouts and things like that. But, as the jobs went away people moved and all the satellite companies that were selling wares to IBM, they just disappeared because with IBM gone their market was gone. It really affected the area and that had nothing to do with the spill. It was just the fact that they were moving out of the area. Basically IBM was picking up and moving out. And of course, EJ was here, but they were almost done when we moved here. Shortly after we moved here, IBM was doing great and EJ started goin down, down, down, down. So it's just one of those things that is happening in many places across the country where you get a big company in a town that moves out, it is just going to change that community big time.

One former IBMer and current Endicott village trustee explained that "We've got Endicotts all over the place." After reminding me that "we live in a classic Rust Belt area," this resident shared with me his perspective on what he feels is the root cause of deindustrialization, economic troubles, and exodus in Endicott and New York State as a whole:

> This is a northern community that was heavily dominated by a manufacturing industry that dates back to the early 1900s, and a lot of that industry has left. Ok. The scars of what that industry was doing have been left behind. Upstate New York is a classic example of this. We've got Endicotts all over the place. To see where industry has left a big scar on the landscape, go to Buffalo's Niagara Falls [he was referring to Love Canal in particular]. They've got real problems up there. They have had tremendous industry up there that just polluted the hell out of the land and walked away. A lot of those industries just don't exist anymore. So a lot of the problems that they left behind are still there and the people are living there. . . . So, what has happened to the economy? Upstate New York is in serious trouble. New York State is very unfriendly to businesses. A lot of heavy taxes and heavy requirements in terms of insurance. Utilities are very high and gas and electric are much higher than the national averages. High taxes. To start a new business here is tough. You see a lot of them start and then fold. Start and fold. Start, fold. For example, one guy had a small business here, and I forget what it was exactly, but he moved just south across the border to Pennsylvania because of the costs. The workmen compensation costs in New York were just way too high. To stay afloat he went across the border. Unfortunately my friends in Albany don't seem to get it. Their solution, like the solution for most politicians, is to increase government spending to fix problems and then our taxes go up. We need to cut government, not increase government. It's a no win scenario. We are in a situation in our state where the taxes go up every year because the cost of labor continues to go up. Unions are very expensive. The costs are going up and everybody is packing their bags and leaving.

This village trustee also informed me that Endicott's budget for FY 2008 was $22 million, which is offset by about $15 million in revenue that comes back to the village in revenues generated from water and electric, parking tickets, and the other $7 million has to be raised in taxes.

Other residents turn to corporate greed and the problem of corporate capitalism when trying to make sense of the causal agents of deindustrialization and political economic transformations at local and national levels. For this resident, whose father was a successful IBMer, the issue is more about a problem of capital being favored over labor, about IBM de-territoriality and broader national and global political-economic trends:

IBM employees are called "Zippers." They are like wind-up toys. Whatever the corporation told them to do, they did. Like wind-up toys who had been programmed, you know. Puppets. Like whatever the company says, the company is right. I don't think much of the corporate world and know it is being proven out nationwide and worldwide that workers are nothing but a number. They really don't matter to the company. And if you mentioned "union" at IBM you were out the door. If you mentioned an IBM union in the 1960s, 70s, or 80s, they would find a way to let you go. . . . Today, everybody is downsizing, cutting jobs and benefits, and squeezing their workers. If you ask for a raise, they will just get rid of you and hire somebody on the street. The outsourcing is happening everywhere and not just here in Endicott. In Michigan, nearly half a million auto-industry jobs were lost. Michigan is in a serious depression and so is Endicott. There is a whole globalization of the workforce. Your job has been outsourced for 70 cents an hour in China. [laughs] Corporations profit from this. Do they lower their prices 100 or 200 percent? No. They drop them a little bit and they grab market share.

He lit a cigarette and continued with his dystopic vision of what this political economic trend is leading to:

I feel sorry for the generation today going into the work force around here that are not educated. I feel sorry for the children. Most of them are educated, but they are going to have problems in the future. I see a future generation in 30 years of people with pensions and people without pensions. There is going to be a big rift where government workers and big companies . . . where you have all the teachers and government workers with pensions and people in the private sector will have their 401Ks and that's it. I see this happening in the next 30 years. I'm not real optimistic about the way things will go.

Shifting Demographics and Emergent Welfare Politics

When discussing the theme of community decline and the local impact of IBM deindustrialization, it was common for residents to talk about Endicott's shifting demographics. For some, with IBM gone and the economy down, post-IBM Endicott is now a welfare community:

Well most of the people that lived here got out of here when IBM started to downsize. Now it really ran down the area. It is all Section 8 housing.[12] The neighborhood has really changed. You know what I mean. You are bringing people in from everywhere. None of them are working. They all got little kids, living on welfare, and because it's free they don't care. You know what I mean. This is a free place to live. Now the property owners like me are stuck paying the bill. You know, where I didn't create this problem. The whole neighborhood has changed. Like I said, we have a Section 8 program going on here. These people are moving here thinking they are out in the country and that it is better than New York City. There are also a lot of absentee landlords. The rental properties are faring well. Landlords are making like $650 to $700 a month for a one- or two-bedroom. It has doubled in the past seven years. Real estate here is all about rentals. It's all Section 8. My big thing is that how do you protect property values[13] when it is now turning Section 8? Someone is pushing the Section 8 stuff. It's something to look into.

With the Section 8 Program going on here, if you wanna have a bunch of kids and don't wanna work, you'll get everything you need here in Broome County. Anything from food to anything. People are moving in here and the place smells a lot better than New York City. All the buildings have been bought up, old buildings that used to have working families. It's all Section 8 now with absentee landlords. The neighborhood has totally changed. I don't wanna live here anymore. The landlords don't care who they rent to, as long as they are getting their rental money. Then the village has to beef up their police force because now you have a bunch of crap. Like I said, the only thing this has benefited is the village. They got a $2.2 million gift. Working homeowners who live with this crap got shit.

One resident spoke of these newcomers, these Section 8 housing occupants, as "less desirable" people, explaining that these newcomers are the result of the "natural" economic changes which lead to "natural" shifts in local demographics:

> Here, IBM moved out so the economy is down so naturally you see, you know, people moving into the area that are less desirable. You know, you couldn't really blame that on to the TCE problem even though that is a problem. It is not the whole thing because you see the economy has changed. With IBM moved out of here, all of the good jobs are gone, so naturally those people have moved on and now the houses are not selling for what the used to. IBM moved out and the tax base of the community fell out.

I would argue that this understanding of newcomers misses two critical issues: (1) it lacks a clear understanding of the intentions of newcomers in general and the intentions of their migration to Endicott in particular, and (2) the fact is that while Endicott's deindustrialized landscape is occupied by different people with different histories, all share the experience of and are witness to the effects of deindustrialization and the larger struggle to live amid neoliberalism (Harvey 2005; Graeber 2010; Bourdieu 2003; Reich 2008).

I interviewed two African American residents who moved to Endicott from New York City, and found that the reason for their migration was not based on some motivation to go on an adventure to find Section 8 housing or because they wanted to leave New York City; rather, their migration story is more complicated. Ron, 45, moved to Endicott from Brooklyn six years ago because his fiancé and daughter moved here to be closer to her family. Clare, 32, moved to Endicott from the south Bronx three years ago with her eight-month-old daughter to be closer to her daughter's father who is serving time in the Broome County Correctional Facility. They ended up in Endicott for these reasons and they are living in the TCE plume with everyone else. Whether one is talking about newcomers employed by Endicott Interconnect Technologies (EIT) earning low wages to manufacture semiconductor packaging, those trying to survive on a monthly welfare check, or old-timers living on a fixed income and left with only their memories of the EJ and IBM boom times, in the end, plume residents are plume residents. Each plume resident is living in the same post-IBM town, the same contaminated Rust Belt community on the wane. Clare and Ron, while newcomers, see the local depression that old-timers are seeing and witnessing, even if from a different socioeconomic or ethnic position. They also blame people who use their welfare check on drugs. In this way, their criticism is compatible with those mentioned previously, even complaints about unfair taxation.

Clare and Ron live in separate apartments in a multi-family home, but "hang out" often. What follows is a narrative snapshot of their Endicott experience as newcomers, as African Americans, as representatives of Endicott's post-IBM demographic shift. For the interview, we sat on the front steps of their building, which were carpeted with artificial turf. It was a sunny and tranquil autumn day:

RON: I am not working now. I have really bad back problems and now my hemorrhoids are bothering me. Social Services ain't helping me out that much. My rent is $350, they pay for that. Then they give me $150 a month for food, but that don't cut it. Especially with the meds I need for health issues.

PETER: How do you like living in Endicott?

RON: This is a good town, as far as getting yourself together. The education is good. As opposed to living in the city, where there are so many people that you can't get that direct help, you know. But, up here you can get that help and if you wanna do it it can happen, you know. I am moving out of here next month. I am by myself and my back is too much. I am going back to Brooklyn next month to be with my family. I will really hate to leave it, you know. I will really hate to leave it. I was really enjoying going back to college as a 45-year-old man. I'm gone. I'm not getting what I need, that's why. I'm not getting what I need, so I am going where I can get it. I am going back to be with my family, so that is what I am gonna do.

If you talk to more people, you know, around my age or ethnic group, they will probably tell you the same thing I'm telling you. It seems like some people get what they get and some people get what they get. You know. It really seems like that. You would not like to think that way, but it seems that way.

PETER: What about you Clare, do you like living in Endicott?

CLARE: I don't like it, period. It's more than just. . . . this whole county is crooked. It's not something that you can't feel. You can feel it, you know. You can also see it. It's like a twilight zone, unless you can hit the lottery, it feels like it is just like digging a hole deeper and deeper. You can also only make so much around, you know, working at EIT or the hospital. With the job market, everybody picks and chooses who they want, so it's like damn you picked somebody else because my skin is a different color and even when you know nothing about me. Even when you go to the super market you can just feel it. I used to work at Price Chopper. You can just feel it. I went to Kmart recently and the lady in front of me was Caucasian and the cashier was sayin' "How are you doing?" then when it was my turn she said nothing and I said "What do you pick and choose who you say hi to?" She was like "Oh, hi" and I was like "It's too late now." It's little things like that, you know what I mean. I don't appreciate it. It's not just one thing. It makes you not want to shop only in certain places, you know. It's all over and it's not like one thing.

It's like the Caucasians out here, when having conversations with them, it's like they have to make a point "Yeah, I have black friends." It is like, what does that matter? It's like they have to make a point to say that to you, you know. I mean it goes from minute to grand. It's just so out there.

PETER: Some people I have spoken to, all white I should add, have said that Endicott is a real welfare community. Do you think this is the case?

CLARE: Well, let's talk about social services. This is my first child and they give hardly nothing [Clare held her sleepy eight-month-old daughter in her arms during the interview]. I have been working since I was 16 years old. The majority of the people out here get it and use it for drugs. They are paying for all of these people who never gonna get a job and live off of my taxes and they are gonna give me a hard way to go. No. I'm not gonna buy. You are gonna give me what I deserve.

RON: It's funny you said that because they just cleared me to work, but I'm still hurt.

CLARE: Uh ha.

RON: So when they send me a letter to go on a job search and I don't go, that means I'm homeless again.

CLARE: I have no family up here and I only know a few people up here and that's by choice. Her daddy was here, but we don't want nothing to do with him no more. Ninety-five percent of the people up here are on drugs or have some type of infection and I'm not interested. I go to the stores and then I come back home. That's it. There is nothing to do in this area. You have to go all the way up the damn hill just to get to a park. They could make some of these empty lots parks or whatever.

RON: Yeah, no doubt.

CLARE: You have got to go all the way up the hill and that is just ridiculous. You know. It's like, I don't know. Then they are gonna raise your taxes, and keep raising them, but what are they doing? They are not doing nothing for us.

RON: The fact is what are we getting out of all this?

CLARE: It ain't like I haven't been working. I worked at Endicott Interconnect Technologies [EIT] when I first came here. I was working for EIT in the "yellow room" where they wash the circuit boards and putting the screws in them. Then they moved me to the plasma department. But in that yellow room there were a whole bunch of contaminants in there. There were times when a coworker would say "Look at my badge" because the badge would be all corroded and stuff. Then there was another incident with a guy who had been working there for 25 years and all of a sudden he came up with a rare blood disease. He could only have gotten it from EIT. So I mean I was there for 18 months and after all that I was through. I said to myself "you know what, my life is more important than any type of settlement you can give me if I'm gonna end up with cancer." Now, yes I smoke cigarettes, but damn, let me get it from that. I don't mean it like that, but I don't wanna get it from working at a job, you know. We would hear a lot of things at work. I mean the money was decent, but it's not worth my health and then you not giving me any benefits or showing me any appreciation as a worker, you know.

RON: They just fire you after six months and just hire a whole new crew.

CLARE: No doubt. That's what they do. And I don't believe the government or these political officials can really help us out. They can barely help us now. We don't have a lot of jobs up here really. If you don't have a car, they just say "to hell which you." You know. So, let's be for real. There is a lot to be concerned about when you are talking about us, when you really put it down in a nutshell. There is not a lot of concern for us up here. You know. It's unfortunate that even in our society I could even say something like that.

RON: In the city, I couldn't even say something like that. Then you come up here and there are less people and more room and you get that. I had a job at Giant and my manager used the N word and I didn't take that too lightly. I worked in the deli department and he said that and I didn't take it too lightly, but I didn't wanna lose my job. It was when I first moved up here. He ended up pushing me out of the job. He wanted to rehire me, but I was not too interested in going to work there. You know. That really hurt me. In the city you wouldn't get that, but up here?

CLARE: It's just not very integrated up here. There is a lot of hostility up here. The majority is still Caucasian.

RON: Don't get me wrong. I got a couple of Caucasian friends. It's real good what you are doing.

CLARE: That's the thing. You don't wanna assume that everybody is like that. You don't wanna put everybody in the same basket. That's why it is so frustrating because sometimes it gets so hard that it is hard not to when it is coming at you from so many different angles. It's not something that I take lightly [she is talking about the use of the N word]. It is different when one of us says it. When a Caucasian says it I don't take it lightly, because of the history behind it. You know. I try not to make everything sound black and white, but damn at the end of it and you frustrate me and you fuck with me, I have to say is it because I am African American that people do this. You got Caucasians out here that get in trouble with the law and get talked to and get a ticket. You get an African American in that situation and they are going to jail as soon as they get in trouble. That's ridiculous.

RON: You hang out with us long enough you'll see what we're talking about.

CLARE: You can't say "oh I empathize or I sympathize." You can't. Unless you are the color of our skin, you will never know what it feels like. You will never know. You can get close to it, but you still won't know. It's a whole different world. I don't want my daughter raised up here or going to the school system up here or none of that. I am trying to move.

RON: Well, on the education part, I disagree which you on that.

CLARE: Yeah, I've heard people say it is good, but there is not enough diversity for me. There is not enough diversity here. I would be lying to myself if I thought my daughter would be getting a fair shot up here. I really think I would be lying to myself and I'm not gonna lie to her. It's a sad situation up here. I don't like it up here at all.

PETER: Because of your experience here, do you think you will move back to the South Bronx sometime?

CLARE: I think about going back down there simply because that is where my roots are and I will be close to my family. Up here I don't have no family, no love. There's no community out here. It's like you are only good to people who want to get something out of you and I don't like people that. I consider myself to be a real generous person and if I have it I don't mind sharing, but if I need don't treat me like, you know, you pissing on my back and telling me it's raining. I don't like that about up here. I get a lot of that up here. It's not so much that I am looking for it, but people are gonna be people. Unfortunately, everybody is not like me. Everybody is not as real as I consider myself to be, you know . . . Your demeanor changes when you move to a different social environment where people are not as considerate. You start to change your demeanor. You know. I don't like that. It's not so much about turning me into a person I don't wanna be, but I can feel that. I think that is where the anger and animosity.

RON: I'm just frustrated. Believe me, I feel it. With my qualifications I feel like I should be able to work in any restaurant here. Any restaurant here, and I had to end up at the KFC on Vestal Ave. Wendy's won't hire me. They said they won't hire anybody who went to BCC. I just can't understand that.

CLARE: It's like if I ain't making the rules I can't create them.

RON: I mean you gotta be diplomatic about things, but sometimes people push you. You understand what I am saying? Who pays the penalty for this landlord not doing what he is supposed to do? It's crazy. When my back went out and I couldn't work, I got an eviction notice. The judge didn't wanna here anything I had to say. He said, are you still in the apartment. I said "Yeah." He said, "That's all I need to know. I want you out by Monday." This was on a Friday it Sunday was Easter Sunday. I had to move out, by myself, and with a thrown out back. There is no way he could override my medical condition. I feel it in my gut. I know it. Nobody wanna tell me where to go with that. Nobody. Somebody needs to be held accountable. You don't just do people like that. I am sure if the judge was hurt, he wouldn't come to work. If he was hurt he wouldn't come to work. He's out of work. I had papers that said I was hurt. It was really wearing me down. Why would you make someone move their stuff? Why would you do that to a person? I get mad just talking about it. He could have just given me a month. I had to leave some of my stuff on the porch. I couldn't carry it. I feel I got played out of that deal. Now the landlord wants $1,500 from me. Nobody can give me answers. I'm gone man.

I mentioned earlier that plume residents, in a way, fold the IBM plume into their identity and sense of place. Each is coping with living in a toxic post-IBM neoliberal world of declining securities. Of course, after listening to the struggle of Ron and Clare in detail, it is safe to conclude that these newcomers of color experience their own particular set of social and economic cruxes. What I find interesting about the preceding narrative is the ways in which, amid their attention to explicit racial tensions experienced in the community, and therefore radical differences in local intersubjective experience, their critiques of taxation and people taking advantage of the welfare system match the critiques of other non-minority plume residents. There seems to be a common *sense of place* (e.g., a place with no jobs and a lack of green space in the plume), but a divergent *sense of community* (e.g., social experience and social ties). But, I am unsure if this is really the case, as only three African American residents completed the survey and I only conducted two in-depth interviews with minority plume residents. But, nonetheless, Ron and Clare helped me to rethink and question the simplistic critiques coming from non-minority plume residents espousing their concerns that the demographics shifts occurring in post-IBM Endicott are resulting in a welfare hot spot of intrusive "others."

Enjoying Endicott and Rolling with the Punches

To say that plume residents feel that Endicott has entirely lost its allure, that the TCE contamination and IBM deindustrialization has totally contaminated peoples' sense of place and security, is surely a mistake. Many residents I spoke with—even those who were highly critical of IBM, the responding government agencies (e.g., NYSDEC, NYSDOH, ATSDR), and the abandonment tendencies of corporate capitalism— still enjoy living in Endicott. "Endicott still has its carrousels and a great high school," as one woman put it. People stick around and "roll with the punches" because this is their home:

> I think a lot of people are stressed these days. I don't think Endicott is the only rough place to live. I still think this is a great country to live in. I would not want to be in Zimbabwe. There is a lot of places I'd rather not be. This is probably better than 10,000 places. Even with the pollution and the small traffic jams we have here, it is still a nice place to live. The future of business here does not look too rosy though. Corporations know that it is too expensive to run a business here. They know that they can just go elsewhere whenever they need to keep up with business . . . You have to have a little bit of sense of humor to survive. I just try to roll with the punches. I like living here. I was born here. I have a sister who still lives here and a brother who does not want to have anything to do with me, but I still like this area. I don't want to live in a big city.

Another resident I interviewed, who was born in Endicott in 1971 and makes a living as a baker, bought his home in the plume in June 2008 for $55,000. Happy with his home and his job, he explained that in addition to the excellent hunting that Broome County has to offer, there are other things that keep him in Endicott, even though he knows the area is suffering from joblessness:

> We are hurting for jobs. EJ was there. Forging Works was here. There was a bunch of them. Of course, IBM was here. They [corporations] just move them over here and move them over there, you know. This is how the businesses work and now Endicott is hurting for jobs. My family is the only thing that keeps me here. My mom basically. She has

emphysema and is not doing good. Basically I just wanted to be close to her. I always grew up here. I know people who stayed here. I moved out of here once, with my first wife, and lived in North Carolina, but this is where I grew up. I plan to pretty much stay in this area.

According to this plume resident, attachment to place is not about economic opportunity, but instead being close to a loved one who needs care. One mother of two boys told me: "My children just know they aren't going to come back to Endicott. There is just nothing here for them." The "boom and bust" of IBM and its abandonment of its birthplace, in this way, has lead to youth disinterest and outmigration. This is impact seems certain, but there are still enduring questions that seem to occupy the minds of Endicott residents, no matter if they live in the plume or not: "How serious is the impact? Who the hell knows? How will we ever really find out?" Asked by a retired IBMer, current Village Trustee, and close friend of a two-time cancer survivor, these thorny questions get to the heart of Endicott's "uncertain" situation. Wisdom coming from lived experience is a guiding source of risk understanding for those dealing with vapor intrusion risk. What seems to matter most is what Ron, who was mentioned earlier, brought to my attention: "I may not know what I am talking about, but I know what I am dealing with." Plume residents are *dealing with* various concerns and stressors, including, but not limited to, racial tensions, unemployment, health problems and, for some, stigmatization.

Several studies on community responses to toxic contamination have found that a common psychosocial response to toxic intrusion is "stigma" (Edelstein 1981, 1984, 1987, 1991, 1992, 1993, 2000; see also Tucker 2000a, 2000b).[14] For many residents I interviewed and spoke with informally, "the plume" is a place, a location or site of significance. The "plume" landscape, as discussed in chapter 2, is now a *mitigation landscape,* a defined space littered with vapor mitigation systems that are like mnemonic devices reminding people the Superfund site they live in. Those living in the TCE plume can "feel" stigmatized: "I can't enjoy my house anymore because of the stigma. Stigma is just the word I use for it, but this is how I feel. It has marked my house. I tell people I live in a toxic dump." The TCE turbulence, as we see here, has created a "mark"—one of the so-called requirements of stigma

(Goffman 1963)—on this resident's sense of self and her sense of her home. People and property, in this sense, share the stigmatization; both are victims of IBM contamination. Not only people and their homes, but even certain streets can be "marked": "It's got a real negative connotation because of this area. You know, anybody that hears where you live in it, it is all like 'You guys live over in the IBM spill?' You don't go around bragging and sayin' 'Yeah, I live over there on McKinnley Ave.'"

One resident I interviewed moved into the plume after being warned of the area's known stigmatism: "I got warnings from my co-workers when I first moved here. There is kind of a stigma that goes along with living here. I kind of felt that when my co-workers started to warn me about the water." The village of Endicott has spent millions since the early 1980s to treat local drinking water. These tests have repeatedly shown to have trace levels of TCE that are well below the levels considered a risk to human health. Despite these efforts, I found that both plume residents and non-plume residents refuse to drink water from the tap.[15] One couple I interviewed told me that they drive more than 20 miles to a spring in Lisle, New York to refill their drinking water. That said, plume living has marked many things—people, property, neighborhoods, indoor air, and drinking water—leaving many stuck with a plethora of question marks to think about and live with.

We have just witnessed the "one-two punch"—IBM deindustrialization and TCE contamination—as experienced by Endicott's post-IBM citizens. Paralleling this "one-two punch" scenario is an effort to soften the blow of TCE contamination with state-led technocratic solutions. The remediation of the IBM-Endicott plume and the mitigation of vapor intrusion have been the focus of a long-term multi-government agency effort. The IBM Superfund Site in Endicott has been termed a site of successful "blanket mitigation" whereby homes in the 300-acre plume have been retrofitted with mitigation systems to divert intruding TCE vapors. How this mitigation effort is understood and experienced by the "mitigated" is the focus of the next chapter.

5

Post-Mitigation Skepticism and Frustration

You will find that the home is the basis of all security.
—George Francis Johnson, local industry legend

They just make so light of everything and assume that these mitigation systems on our homes are fixing everything. I don't believe it.
—Plume resident and wife of two-time cancer survivor

While the majority of my fieldwork took place in Endicott, there was a "multisited" ethnographic component to the research (Marcus 1995). In early October 2008, I attended a two-day classroom training entitled "Vapor Intrusion Pathway: A Practical Guide" in Portland, Oregon, that was organized by the Interstate Technology and Regulatory Council (ITRC). Established in 1995, the ITRC is a state-led and government-funded[1] coalition made up of state and federal regulators, industry representatives, academics, and stakeholders working together "to achieve regulatory acceptance of innovative environmental technologies and approaches" (ITRC 2007). In this training I was told by one of the instructors that Endicott was a good example of a site where the decision to mitigate was clear, given specific site characteristics that included, but were not limited to, the extent of detectable TCE levels and the relatively shallow distance between the contaminated groundwater and the slabs of homes. In other words, "blanket mitigation"—the term commonly employed for decisions leading to the mitigation of *all* homes within a zone of contamination with a known or expected threat of vapor intrusion—made sense in Endicott. As he put it, "Some mitigation decisions are a slam dunk. You know they need to be mitigated." Endicott is one of two well-known sites—the other is called the Redfield site in Colorado—to result in a "blanket mitigation" decision. But, as vapor intrusion scientists told me, pre-emptive mitigation does not mean that monitoring is no longer necessary. Post-mitigation monitoring generally involves monitoring the mitigation technology (e.g., sub-slab depressurization system) to see if it is functioning properly. In rare cases, post-mitigation monitoring can involve soil gas sampling and indoor air tests. But, most of these tests cost money, and so monitoring the mitigation system is the most cost-effective option. What is not included in the practice of post-mitigation monitoring is the analysis of the social impact of vapor intrusion mitigation decisions. This "other" or non-traditional monitoring research is a primary focus of this book and is influenced by the very basic idea that, like power relations themselves, mitigation politics need to be demonstrated, not assumed (Wolf 1999).

For this chapter, I draw on narratives from in-depth ethnographic interviews and quantitative survey data to explore local understandings of the mitigation effort and the lived experience of residents living in "mitigated" homes in the IBM-Endicott plume. In other words, my aim is to better understand the social context of life mitigated. While many regulators, vapor intrusion scientists, and even TCE and vapor intrusion activists consider mitigation a righteous decision when a lack of knowledge of risk is available, it is important to note that "blanket" mitigation in Endicott did not lead to a reclassification of Endicott's Superfund status. The IBM-Endicott site is still listed on New York State's Inactive Hazardous Waste Site registry as a Class 2 site, meaning that the TCE plume still poses a significant threat to human health and the environment, as it was first determined in 1984.[2] It also has not led to the "mitigation" of local TCE risk perception, as will be shown later. When it comes to discussions of risk, it is more often than not the case that "there is no scientific monopoly, since there is rarely expert agreement either on what constitutes acceptable risk, or on how it may be managed" (Caplan 2000:3). What results from this situation is a social experience of TCE risk that is very similar to the social experience of

TCE risk mitigation and the possibility of the proliferation of "public criticism and disquiet" (ibid.) even amid what might be considered by many experts and activists a laudable risk management decision.

What follows is some background on the state of the practice of vapor intrusion mitigation and the ways in which "pro-mitigation" discourse plays out among regulators and activists alike. I analyze plume residents' understandings of mitigation and the experience of living in a "mitigated" home, focusing in particular on the theme of persistent or durable ambiguity that we might think would have been softened by IBM and NYSDEC mitigation efforts. Another goal here is to engage a topic that has become a popular source of concern for communities threatened by vapor intrusion risk: property devaluation. I analyze this issue of property devaluation as an example of what I call "mitigation nuisance," or a remedial decision or "treatment" that introduces new forms of distress for homeowners. On the other hand, property devaluation is less a concern for renters, who are a growing population in Endicott's plume. Therefore, a final goal of this chapter is to expose an issue that I argue exacerbates the "political" ecology of risk and mitigation unfolding in Endicott, which is tenant notification politics and using ethnography to measure the effectiveness, or lack thereof, of the tenant notification bill mentioned in previous chapters. This law was designed to inform renters about the plume and most important to provide them with results from air sampling data, if the data for the rental space are available.[3]

Vapor Intrusion Mitigation

At the heart of this book is a desire to better understand how residents of Endicott's TCE plume make sense of mitigation. As pointed out in chapter 2, Endicott was once referred to as a "valley of opportunity," a Big Blue landscape that at its peak was inhabited by 14,000 IBMers. Today, however, Endicott is a "valley of mitigation," the home of the nation's largest concentration of vapor mitigation systems. These mitigation technologies were a "gift" from IBM, and the NYSDEC and other vapor intrusion regulators and scientists call the Endicott Superfund site a good example of "blanket mitigation," whereby the decision to mitigate is based on site-specific considerations that usually call for long-term measures (e.g., mitigating 500 homes) because the complete removal of TCE from the groundwater source is considered technologically infeasible. In many ways, what cannot be fully remediated is consigned to mitigation, and this is how the mitigation decision-making process unfolded in Endicott.

The vapor mitigation systems (VMS) used in Endicott are the same mitigation technologies used to mitigate radon gas, which results from the natural decay of uranium found in the earth's soil and water. VMSs (or sub-slab depressurization systems) are applicable for slab-on-grade building construction. VMSs are technologies designed to continuously lower pressure directly underneath a building floor relative to the pressure within the building. The resulting sub-slab negative pressure inhibits soil gases from flowing into the building, thus reducing, in the case of Endicott, TCE vapor entry into the building. The TCE vapors caught in this negative pressure field are collected and redirected into the ambient air (or the air outside the building). The depressurization under the slab is typically accomplished with a motorized blower. The VMSs in the IBM-Endicott plume have a 90-watt motor that runs 24/7. The blower draws air from the soil beneath the building and discharges it to the atmosphere through a series of collection and discharge pipes. The U.S. EPA (1994) defines these mitigation technologies as "a system designed to achieve lower sub-slab air pressure relative to indoor air pressure by use of a blower-powered vent drawing air from beneath the slab." To know whether or not depressurization is occurring, the VMS has a pressure gauge. One NYSDEC official I spoke with pointed out that the pressure gauge "shows you that mitigation is working. You can just look at the gauge and you can know." In this way, the pressure gauge not only allows for self-evaluation, whereby each mitigated resident can assess the efficacy of mitigation, but it has a concrete techno-science risk communication function: the gauge is right, your response is your problem.

But there is another twist to the vapor intrusion mitigation story. Mitigation is considered by activists, vapor intrusion scientists, and regulators to be a righteous vapor intrusion risk management decision. None of these groups of stakeholders tend to question the efficacy of these technologies to mitigate vapor intrusion risk, arguing that VMSs have helped to achieve the very low action levels required for indoor air remediation. Surely, if properly installed, these technologies can mitigate the effects of TCE vapor intrusion and achieve indoor air action levels, levels of which have been ardently contested in states like New York that have or had a productive industrial sector.[4] Sometimes they can even mitigate intruding volatile organic compounds (VOCs) to levels that are actually non-detectable with air samplers used to measure vapor intrusion.[5]

What follows is the perspective of one NYSDEC scientist and vapor intrusion expert working on the IBM-Endicott site who believes the mitigation decision has "kept people from being exposed" despite a clear technical challenge of fully remediating the TCE plume. His detailed knowledge of the history and particularities of the IBM-Endicott site mitigation decision are informative:

> The two key things as far as the IBM issues specifically, is that we fairly quickly identified the structures that were potentially impacted and put mitigation systems on. I mean we started sampling in the beginning of 2003 and by that summer I think we had probably identified 85 or 90 percent of the homes that we ultimately offered mitigation systems to. We continued to do sampling for the next couple of years, but you know there was a huge effort to get our arms around the problem as quickly as possible. People worked very hard to do that. I did and folks here did, as well as IBM. So since then the emphasis has really been on, ok, we have kind of put the band aids on, we have kept people from being exposed, now what can we do to reduce the source of this contamination, to clean the groundwater up, both on-site and off-site. We have been making, I think, substantial progress in that effort. . . . We have pretty much isolated the groundwater on-site and improved the containment on-site, and we are seeing concentrations decreasing in the groundwater off-site. There is probably a point at which you reach some sort of point where, I mean, like the cleaner it gets the more difficult it is to get cleaner. At some point there is a fairly limited amount of mass left and it can be in places where it is hard to get out. So that is a technical challenge in that respect, but we have actually accomplished quite a bit that way and expect to make progress on that.

During my interview with this NYSDEC official, he also ensured me that the VMSs work, joking that "we don't need a new theory of physics. They work." I explained to him that many residents of the plume that I interviewed complain that nobody has retested the indoor air to assure them that the mitigation system is working. After sharing this finding with him, he responded:

Well, by and large, once we put a system on we do test it physically. I think that is the most important thing because what we are talking about with these systems is changing the physics of flow of contaminants beneath the house. It's about capturing those contaminants in a suction system before they can migrate into the house. So if you put a system in properly and you show through the physical testing that it's in properly, that's a pretty compelling line of evidence that the system is working. Because there were so many homes mitigated in Endicott we sampled only about ten percent, about 45 or 50 homes, of them over a range of concentrations that were in the indoor air. We sampled these homes after they had been mitigated and those data are also compelling.[6] The chemical data basically substantiated the physical data that these things do reduce the concentrations. When we look at it and our management looks at it, those tests are not cheap. I mean it costs about $1000 a test. When we have compelling evidence that says the house has basically been properly mitigated, we say that there is probably a better way to spend it, like maybe looking for other houses. Again, you know, I don't think there is a sound technical basis for going back on a regular basis to test chemically these houses.

In fact, I co-authored an article recently on this issue. The other author of the article has sampled homes at his site [the Redfield site in Denver, Colorado][7] every quarter for like 8 years and there is really no change. The data continue to show that the systems are working. So we are not looking only at data collected in New York, but data that has been collected elsewhere. All the data points in that direction.

As will be explored in more detail later, all the data in my ethnographic findings *point to* an omnipresent sense of ambiguity and precariousness among residents living in mitigated homes. This NYSDEC official explained that he understands the frustration that residents go through, but he is also "comfortable" knowing that the mitigation technology works, which puts him in a sticky position: "I understand that people want certainty. I am pretty comfortable with the way that we install the systems and test them afterwards physically, and again, some chemically. It just becomes a question of, you know, is it a good use of resources to continue to prove that what you did is good."

In Endicott and beyond, community stakeholders engaged in TCE activism and vapor intrusion activism are strong promoters of mitigating homes in areas at risk of vapor intrusion. As Lenny Siegel, co-founder of the Center for Public Environmental Oversight, community stakeholder, and technical advisor to communities impacted by vapor intrusion, puts it, "Fortunately, it is relatively easy and inexpensive to prevent vapor intrusion. Sub-slab and sub-membrane depressurization systems, developed through decades of response to radon intrusion, can prevent the flow of contaminants from the subsurface into buildings" (Siegel 2009:6). Siegel warns that although mitigation systems like those installed in Endicott can be an effective form of vapor intrusion mitigation, *"they only work as long as they work.* To ensure that building occupants are protected, mitigation should be anchored in long-term management, which includes operation and maintenance, monitoring and inspection, contingency planning, notification, institutional controls, and periodic review" (2009:8; emphasis in original).

Vapor intrusion activists and community stakeholders believe in this pro-mitigation effort and point to the fact that mitigation decisions are often most contentious because VMSs are offered to some residents and not others based on the boundaries and concentrations of the VOC contamination. In one stakeholder report (Siegel 2008), one activist from a vapor intrusion site in Dutchess County, New York contended that

> I think any home that is between two other homes needing mitigation should be mitigated. I have come to that conclusion by putting myself in that person's place. How would I feel knowing that all around me the air is contaminated and my home is not protected should the toxic vapors decide to invade my home? I feel the same way for homes that are considered just outside the contaminated [zone]. . . . The only way to breathe easy is to clean the plume that is the cause of the contamination. . . . I still do not understand why agencies decide to monitor and not mitigate should they find contamination since in many cases it costs less to mitigate than it does to monitor. (Siegel 2008:5–7)

This last point about the cost-effectiveness of mitigation is important. According to the U.S. EPA, installation costs for active venting systems like the ones populating the Endicott landscape range from $1,500–$5,000, and the cost of annual operations and maintenance can range from $50–$400 (USEPA 2008). Continued air monitoring and lab costs

Table 5.1. Sub-slab Depressurization Systems: Pros and Cons

Pros	Cons
Successful track record of performance, 90–99% reductions typical, 99.5% or greater reduction possible with well-designed systems	Requires periodic maintenance
Adaptable technology, applicable to a wide variety of site conditions and geology	Wet and low-permeability soils retard soil gas movement
Simple gauges show whether the system is working	Building-specific conditions may limit options for suction pit, riser pipe, and fan locations

*Adapted from ITRC (2007:51).

for analysis per home can range from $8,000–$12,000 and can vary with monitoring duration. There is general agreement among environmental scientists and engineers that mitigation technologies do in fact do a good job of mitigating vapors, but to maintain the "control" of vapor intrusion these systems do require periodic maintenance and can't mitigate 100 percent of all volatized organics in the indoor air. In some instances of vapor intrusion, there are chemicals of concern that might affect the mitigation system design and exacerbate the risk scenario. For example, during an interview with Kevin, a "mitigated" resident, he brought me down to the basement to show me an alarm system that was connected to his mitigation system. Kevin said he didn't really know why he had this alarm system on it and was not sure if other people had it too. I told him I had never seen such a thing, but later found out that when potentially combustible vapors or those that may approach combustible concentrations are found during pre-mitigation sampling, it is common to install an alarm system on the mitigation system. After learning this, I called Kevin to inform him of my findings. From what I could tell, he appreciated the call.

Again, vocal residents and community stakeholders see mitigation as a laudable protective decision. One vapor intrusion activist, who happens to also be a professor of geology and is a founding member of the New York State Vapor Intrusion Alliance, was also quoted in this same stakeholders' report (Siegel 2007), saying that

Under New York State's guidelines, if three homes are located on a toxic plume and the outer homes qualify for mitigation, VOC indoor levels in the middle home may not be high enough during a test to warrant mitigation. The middle homeowner has to suffer unacceptable anxiety during the period between tests to see if the levels may rise to the action level for mitigation. Therefore, to reduce anxiety of residents in a "monitor" phase, any home with vapor intrusion from an outside source should be mitigated . . . [He and his advocacy group] strongly supports blanket mitigation. All homes, schools and businesses on a known VOC plume should be mitigated to ensure against future intrusions of these unwanted toxins.

Active sub-slab depressurization systems—what I am calling "vapor mitigation systems"—are no doubt the "most widely applied and effective systems for vapor intrusion control" (ITRC 2007:50). They are "widely considered the most practical vapor intrusion mitigation strategy for most existing and new structures, including those with basement slabs or slab-on-grade foundations" (ibid.; see also USEPA 1993). But, what do residents living in mitigated homes have to say about mitigation? How is mitigation experienced and grounded by real people living in an environment of mitigation? There is another world outside the vapor intrusion science and data supporting the argument that these systems are efficacious, that they control intruding TCE vapors.

Life Mitigated: Ethnographic Traces

The medical anthropologist Arthur Kleinman (1980) introduced the concept of "explanatory models" to explain how people make sense of their illness experience (e.g., illness causation, diagnostic criteria, and treatment options). For Kleinman (1980), a focus on explanatory models was critical to better understanding the reality of the clinical encounter, because the explanatory models held by medical practitioners, patients, and family can, and often do, differ. This same logic applies to vapor intrusion mitigation, and what follows is my attempt to draw on interview and survey data to expose what might be called "explanatory models" of mitigation.

In many ways this book aims to ground the experience of mitigation in ethnographic findings, because I feel that those living outside of the IBM-Endicott plume misrepresent the experience of those living in the plume and living *with* mitigation. For example, I interviewed an Endicott resident who lives outside of the plume. His home does not have a VMS, but this doesn't stop him from having an opinion about the efficacy of the VMSs.

For him, they will be "working" in Endicott for a while: "They [IBM] are going to need to run those things for some time to come, it would appear. I mean they are working. I guess there is sometimes a question of what happens when the fan stops. I guess overall, these ventilation systems are necessary and keeping the people at least safer in their homes."

Like the NYSDEC official mentioned earlier, this non-plume resident is sure that mitigation is working. This is not to say it is working for the mitigated, if "working" means these technologies have in fact mitigated plume residents' sense of risk and uncertainty. Rather than eliminating fear or extinguishing residents' sense of risk, almost every mitigated resident I interviewed questioned or was perplexed by the possible risk of the toxic vapor being dispersed from the VMS pipe above the home. For Susan, who has lived in the plume since 1985, uncertainty endures, despite her own sense that IBM made a good effort to do the right thing:

> Well I thought it was good they came through and put these pipes in, although I do wonder where all this stuff is going. The pipes go to the top of the roof and then what? Where is it going? It's going right into the air. When our children were young and we would go outside, I could smell something. It really was quite a strong smell. At that time there was nothing about the pollution they were giving out, but I can remember the smoke stacks and everything. I don't really know what is going on. It just seems like a lot of it must be going into the air and we gotta be breathing it if it is going into the air.

In 2006, the Agency for Toxic Substances and Disease Registry (ATSDR) conducted an ambient air study to address community concerns about ambient air issues. The ATSDR report, which analyzed historic emissions data (1987–1993) from the former IBM facility, determined that three chemicals of concern—formaldehyde, methylene chloride, and tetrachloroethylene—needed further evaluation. ATSDR also found that TCE was a concern because of its presence in other environmental media (e.g., indoor air) and because of community concerns and information gathered during ATSDR's evaluation process. ATSDR evaluated the possible health effects of past air exposures to these four contaminants and determined that these past exposures present "no apparent public health hazard" (ATSDR 2006:3). What this claim means is that

adverse non-cancer health effects are not "expected," and that the likelihood of cancer resulting from an exposure during 1987–1993 was "very
low to low" (or ranging from greater than one theoretical excess cancer case for every million persons exposed to less than one theoretical excess cancer case for every 10,000 persons exposed).[8] ATSDR also
evaluated the possible health effects of exposures to chemical mixtures,
or the combination of multiple VOCs, finding that "adverse non-cancer
health effects are not expected, and the cancer risk from the combined
past air exposures to VOCs is considered to be low" (2006:3).

Despite ATSDR's study and hard work to determine "low" risk, many
plume residents are uncertain about the ambient air issue, and even
contend that the VMSs are exacerbating the ambient air impact. One
resident I interviewed added that it is kind of "strange" to assume that
sucking vapors from underneath the home and releasing them above
the home is a risk-free process, because "no one knows for sure":

> Well you know this system is taking the vapors out of the ground and
> it is the same system used for radon. It is the same idea. As the vapors
> come through the ground, instead of rising through the floor and what-
> ever, they are sucked out into the air. On our property they put one on the
> crawl space and one in the basement. So there are two systems. Of course,
> that was one of the very first questions we asked. Ok, you are taking these
> vapors out of the ground and you are putting them in the air. Initially they
> told us "Oh no, as soon as it hits the air it's immediately dissipated." We
> thought "well that sounds kinda strange." So, now that is kinda up in the
> air. But assuming that system is working, we have not been tested. We
> requested to be tested, but they said "No, it is not necessary because we
> are putting this system on everybody's house so you don't need it." They
> never tested it in the first place, so how do we know it is working? No one
> knows for sure that by pumping this stuff into the air it really totally dis-
> sipates. We wonder "Does it truly dissipate?" Then the discussion is back
> and forth. "Now the air quality is bad because of it." There has really been
> no conclusive test, at least that I know of, showing one way or the other.

Others explained this ambient air risk situation this way: "Here it is for
me: All that stuff is being sucked up and then it is going into the air and
everybody is breathing it. If there is no wind, ok, then what?" Or, "They

say it dissipates, but who really knows." Another offered a technical solution: "It baffles me that they did not put carbon filters on the end of those things." When asked if she felt having a VMS installed would fix the problem, this old-timer responded:

> Absolutely not. I'm a skeptic from way back. I'm Irish [laugh]. One thing I asked them was "Ok, these things are venting it out. They are sucking it out from the basement, bringing it up through the pipe, and it goes out. Where does it go?" I mean it's out. It's the same stuff that was down there, but now it's here. Now they just throw it up.

Aside from this common critique of mitigation as a process of re-directing the TCE vapor, rather than really mitigating TCE from the plume environment, there is a local *doxa*[9] regarding mitigation, which Tom, a retired IBMer, summed up nicely:

> Is it actually doing something? I guess it is because they wouldn't have wasted all their money, you know, putting them in if they weren't doing something. But, is the effects of it, you know, that good? We don't know. You know, at least if you saw something. It seems to me like any filtering system or whatever. If you saw, you know, residue, or if you saw a discol-oration that would make you realize that something is coming out. But, we never see anything. There is no filter to see or nothing obvious. All we know is that that motor is running. I can figure that out for myself.

These narrative findings match up with much of the uncertainty found in the quantitative survey results. My survey of mitigated plume residents included the statement, "The VMS has reduced the TCE levels in my home," and asked residents to rank the degree to which they agreed or disagreed with this statement. The majority of residents surveyed (n=55, or 67%) to answer this question marked "Don't Know." After hearing how important mitigation was and learning that both regulators and VI activists were gen-erally pro-mitigation, I found this result surprising. Many residents I inter-viewed seemed to be frustrated that when they requested monitoring after the VMS was installed, all the inspection process included was a motor check and inspection to see that depressurization was occurring. Many residents seem to be led to believe that monitoring meant retesting indoor air concentrations to compare pre- and post-mitigation levels. This was not the case. As the NYSDEC official mentioned earlier put it, "These systems work. All the data points in that direction." But, if this is the case, I wondered, why such a strong Socratic tendency ("I know that I don't know") amid what many would consider a praiseworthy mitigation effort?

While it is common to discard "Don't know" responses when analyzing survey data, I found this finding important in its own right, especially as a measure of the shortcomings of an intervention designed to mitigate and control vapor intrusion and, therefore, community concerns. Ultimately, this survey result highlights the politics of uncertainty for residents of Endicott's IBM plume. It is this interface of mitigation science, the state, and the perplexed citizen that I think sheds light on the utility of the political ecology perspective, which ultimately, in this vapor intrusion mitigation setting, is about articulating the ways in which impacted residents are engaging the politics of mitigation and immitigability. Following West (2005), I think it is the "transactional being-in-the-world, in which subjectivity is constantly being produced" (West 2005:638–39) that ought to be the central focus of an environmental anthropology of vapor intrusion mitigation. As discussed in chapter 1, the political ecology of mitigation this book encourages is one that describes and shows how vapor intrusion mitigation understanding and experience is "aesthetic, poetic, and deeply social" (West 2005: 639), not just how mitigating vapor intrusion in Endicott is connected to and informed by shifting and stabilizing environmental science and policy at the state and national levels. "Not knowing" the efficacy of mitigation is a significant finding, as are the daily nuisances of mitigation technology.

Mitigation as Nuisance

What I shall call "mitigation nuisance" is really the unexpected consequences and concerns that result from mitigation decisions, like the "blanket mitigation" decision that has turned Endicott into a mitigation landscape colonized by more than 500 VMSs. One iatrogenic effect of mitigation, for example, is that homeowners have to live with the "eyesore," as one resident put it, that comes with vapor intrusion mitigation. For Shannon, "It looks ugly. That is why I had them put it behind the chimney. I didn't want people to see it." Others point out the noise that the VMS makes. The 90-watt motor running the VMSs on homes in Endicott's IBM plume runs 24 hours a day, seven days a week. The VMS makes a humming noise that can irritate residents. Shannon explained it this way: "That motor is loud. We hear it all through the night. I had to move my China closet because it was against the wall with the motor and the glass doors on it kept clanking. So, what is it doing? I can tell you the motors are running [Shannon has two VMS units on her home]. The noise reminds you it's there. I wasn't too happy about having a ventilator on my house because when they put it on the house the house hummed."

Shannon even had IBM move it away from the wall so it was not flush against the exterior wall, which was what created an annoying rattle. She contemplated keeping it the way it was, because she like that it was hidden and tucked behind the chimney, but the noise eventually got to her. She tested her patience for a couple of months after it was installed and then called to have it moved after she realized she could not live with the noise. Shannon's noisy mitigation system indexes a difficult experience of what philosopher Judith Butler would call "up againstness" (Butler 2011), whereby mitigation technology, the home, and Shannon are caught up in precarious cohabitation or uncomfortable closeness. Coping with the circumstances, for her it came down to a choice between picking an "eyesore and the noise." Shannon chose the former. According to one NYSDEC official I spoke with, of 401 mitigation systems in Endicott in 2008, ten noisy fans were reported.

Some residents live with the noise and point out that it just starts to go away once you stop focusing on it. On the other hand, this trick only lasts so long. For example, plume resident of 40 years, Lucille told me: "It vibrates and granted sometimes during the day you just get used to it and don't even hear it. But, at night, sometimes, and especially with all the kids gone, I will be sitting by myself and turn off the T.V. and be writing or something and I hear that noise. If I go up stairs and go in the room where I do my crafts late at night, that motor sound is there. You can hear that motor going." In many ways, the VMS is like a mnemonic device, a constant reminder that the plume lives on, even when you would rather forget and wish it was mitigated and gone. It should be pointed out here that upon closer inspection of the etymology of the concept of *mitigation*, it is not about getting rid of or extinguishing something, but instead "softening" the presence of something that, like TCE vapor, is bothersome or intrusive.

Figure 5.1 Shannon's Noisy Mitigation System. Photo by Peter C. Little.

To correct the eyesore effect of the VMS, some residents try to blend the VMS in by coloring the white piping. On a gloomy Saturday afternoon during one of my plume walks, I came across one resident who had put painting the VMS on his weekend "to-do list." I ran into him just as he was finishing. We chatted briefly in his driveway, where he explained that "the white looked ugly" and that he was also painting it because he was planning to try to sell the house in the next couple of months. Several churches in the plume have also made efforts to blend in the VMS. The First Baptist Church, located in the heart of the plume, painted the VMS piping to blend in with the color of the brick.

Property Devaluation

Communities affected by vapor intrusion tend to worry about two primary issues: health impacts and property devaluation. The latter is of particular interest here because even efforts to mitigate vapor intrusion mark a structure as a source or site of vapor intrusion, even if the home has been determined by science and the state to be a "mitigated" structure. Mitigation systems, like the presence of contamination itself, leads to "the inversion of [the] home" (Edelstein 2004:93): "The meaning of home as haven from a complex society is inverted by toxic exposure" (ibid.).[10] The same can be said even when homes have been mitigated and recent social complexities and crises have contributed to this process of inversion. For example, the meaning of home as a good, lifelong investment has been inverted, due largely to the real estate market crash of 2008, which was in full swing just as I began my fieldwork in the summer of 2008. While one newcomer to the community told me that her realtor said that the VMS was a "home improvement," this did not seem to come up in my interviews with plume residents. Instead, homeowners seem to be collectively concerned about property devaluation and frustrated about paying taxes on a home tainted and devalued by IBM TCE contamination. What follows is just a snapshot of some of the narratives I collected on this theme.

Property devaluation was a highly contentious theme to surface in my interviews with homeowners; the majority of homeowners who responded to the survey (n=50) either said they were "very concerned" (29, or 58%) or "concerned" (13, or 26%) that the IBM contamination had depreciated the value of their home. This property devaluation concern is exacerbated by tax equity politics, with many homeowners in the plume contesting the property taxes they owe Endicott.[11] Most of these residents felt they should not be paying the same property taxes that non-plume residents pay. Much of the taxation talk turns to a critique of IBM—sometimes explicit, sometimes implicit—and IBM's abandonment of Endicott and leaving Endicott with a massively reduced tax base. In other words, the connections between property devaluation, taxation, and IBM deindustrialization were sharply defined. In the word of one plume resident, retired IBMer, and proprietor: "The whole village just took a really hard hit from all of this. With the taxes and everything, it's just a downward spiral."

One former IBMer and village board member empathizes with plume residents concerned about property devaluation, arguing that IBM should have offered plume residents more than they did to cover the property value damages, damages that VMSs don't get rid of:

> Maybe IBM should have offered more than $10,000[12] because of the property devaluation issue. . . . It's an enormous problem. People who live in the plume have a problem. Even if the property looks like it is going fine and the mitigation system is working fine and there is nothing wrong with the property, just the fact that they live in the plume is killing the value of their property. If the plume wasn't there, everybody would be fine. They got screwed. A lot of people packed up their bags and moved out of that area. Some people

can't afford to leave or can't afford to take the loss of selling their home for nothing and moving someplace else. A lot of people are just stuck.

Being "stuck" in the "mitigated" plume as a homeowner is a serious problem that surfaced in one of my interviews with a couple who have been living in Endicott since the mid-1980s. Their story and situation are not unlike that of other homeowners living in mitigated homes with declined value:

LOU: The value of our house has gone down the tubes. You know, we tried to get it appraised and we couldn't get it appraised for what we bought it for . . . So I think one of the things that really bothered me is the property value. Even when we tried to get the house appraised we knew that we had probably put about $20,000 into remodeling, changing and replacing all the windows, we built this back deck, and yet when they come they can't even appraise it for as much as we paid for it.

TIFFANY: They just turned us down outright. They could not give us enough for the mortgage and finally FHA did for $1,500 more than we paid for it 22 years ago. I feel and I think my neighbors feel too that, I mean, because of this there is no way we can sell our homes for what we think they are worth. They will laugh at that too. They will say "Everybody thinks their house is worth more." I can show you bills. We put in $20,000 and you mean to tell me, you know, that I only appreciated $1,500 in over 20 years. I mean, so now its kinda of a joke and I think a lot of my neighbors feel the same way. You know, somebody will say "Where do you live?" "On the avenue in the plume" [laugh]. I mean, come on. Our property value has depreciated so much.

LOU: Yeah, so it's got a real negative connotation because of this area. You know, anybody that hears where you live it's "Aw, you guys live over in the IBM spill." Now we have not tried to sell our house, but with the way the appraisal process went. Well we are just lucky to get a refinance, but ah, it took some real effort on our part . . . Unfortunately, you know, it's like the damage has been done and I don't know if, as far as property value goes, people will ever want to buy a house in an area that they know has been polluted. They know they can just buy elsewhere. Here it is a question mark. We have been here 20 years and things have just gone downhill around here. Can we blame this all on the TCE? If

I did sell, I would be taking a huge loss. Two houses on this side of the street have not been able to sell and one 4 bedroom house across the street sold for $38,000. That wouldn't even cover my mortgage if I had to sell at that price. I wish IBM would just buy my house, but I don't see that happening. Will IBM step up to the plate and really do that? Well, IBM did offer that $10,000. With that they are saying that they are paying for the damages to the property value. Had I known at the time that it was not taxable, I probably would have taken it because my problem is that my son had just started his second year of college when this happened. If I took the money, it would have increased our reported income and so would have impacted my son's ability to get financial aid. Also at that time I was just starting to collect social security, so I was in that magical 62–65 age. So anything over $10,000 a year I have to give back. It's like for every $2 you make you have to give back $1. I don't know if I would have taken it or not anyway, but I didn't.

Like many plume residents who were offered IBM's $10,000 for property damages, Tiffany and Lou turned it down, even though they might not have, had they known it was not taxable. Of the homeowners I surveyed who were offered IBM's $10,000 settlement (n=38), only 14, or 37% took it, and the majority (24, or 63%) declined the offer.

As pointed out earlier, one resident told me that her realtor referred to the VMS as a "home improvement."[13] While this might be true, many homeowners I interviewed are reluctant to invest in home improvements and don't believe a mitigation system is a remedy for trenchant property devaluation. Another couple I interviewed had this to say:

DON: Our main concern is the property value. . . . I certainly think the property value has gone down here because we are right here in the middle of this crap. That is what I call it. I call the plume the crap . . . We bought this home in 1970 for $18,000. We put money into it, but we will never get out of it what we put into it. . . . We won't do anymore upgrading to the house. It's just not worth it. I won't do nothing.

MARY: We just don't feel it makes any sense to put any more money into this house.

DON: Nothing. She wants a new kitchen, but we are not going to get involved in that.

Many residents of vapor intrusion sites have reported that exposing information about contamination drives down home values, but property taxes stay the same. While this problem may influence a homeowner's decision to not invest in remodeling their home, they can also be reluctant about vapor intrusion testing itself. For example, it has been found that "homeowners are often reluctant to allow outsiders to collect samples on their property" (Siegel 2012:3), though I did not witness this in Endicott. In fact, I actually found the opposite to be true. "Mitigated" residents actually welcomed indoor testing, but the state determined it was unnecessary in a post-mitigation context. Indoor air testing after mitigation was not mandated, nor considered scientifically necessary because sub-slab depressurization technologies "work," as several vapor intrusion scientists and regulators told me. While motivations for home improvement were at a low, desire for post-mitigation indoor air testing was a high priority of plume residents and considered by many a good way to improve residents' understanding of their actual risks of exposure to intruding TCE vapors. Furthermore, the play of certainty and uncertainty engulfs the mitigation landscape. Property devaluation and community corrosion was a *certain* reality for most residents, while the efficacy of risk mitigation and the science of vapor intrusion risk remained *uncertain*.

Overall, the above narratives illustrate some of the lived experiences of the "inversion" (Edelstein 2004) or subversion (Douglas 1993) of the home thesis. Frustration and insecurity persist amid IBM and NYSDEC remediation and mitigation efforts. No matter how much the plume shrinks as a result of the NYSDEC's aggressive groundwater remediation project or how well the VMSs work to reduce TCE concentrations in indoor air, the fact is that property devaluation, the "empire of risk" (Ferguson 2009) resulting from the real estate market crash and financial crisis of 2008–2009,[14] and national and global social and economic insecurity seem to be as intrusive as the risks of TCE vapor intrusion.

Tenant Notification Politics

The political ecology of mitigation perspective proposed at the beginning of this book, and made explicit in this chapter in particular, is strengthened by a committed effort to draw on ethnographic description to ground local mitigation discourse and make the voice of

Endicott's mitigated citizens more audible and meaningful. But this political ecology of mitigation (or PEM) perspective loses strength or influence if it does not attend to people commonly left out of the debate in Endicott, and that group of people includes renters living in the plume. An almost unspoken issue at the local level in Endicott's plume is the politics of understanding and knowledge between home-owners and renters, the latter group making up a large portion of the IBM-Endicott plume population. While state Assemblywoman Donna Lupardo was instrumental in getting New York's Governor David Pat-terson to sign into law the tenant notification bill (A10952-B) requir-ing landlords to notify tenants of the results of environmental testing, to my knowledge this law is rarely, if ever, enforced, leaving renters in a situation of not knowing about the plume they live in, nor what the VMS on their rental property is up to. For me, this is what the political ecology of mitigation at the local level is all about (Little 2013a).

One of the more statistically significant results of my survey of plume residents was that owners and renters living in homes with mitigation systems have differing levels of understanding when it comes to know-ing how the VMS works. When asked to rank, based on a scale from 1 to 5, with 1 being "Strongly agree" and 5 being "Strongly disagree," the survey question "I understand how the VMS Works," renters were much more likely to express that they did not understand how the VMS works.

According to the Interstate Technology and Regulatory Coun-cil (ITRC), community stakeholders at vapor intrusion sites have demanded that "mitigation techniques . . . should include a long-term monitoring plan and a contingency plan" (ITRC 2007:A-3). What my community survey results indicate is that what also ought to be included in this "monitoring" plan is a greater effort to inform renters who are not being informed about vapor intrusion mitigation. Homeowners tend to dominate the pool of "community stakeholders" involved in vapor intrusion debates, and the renter perspective and experience gets lost.

In short, the lesson learned is that we should not assume that the more people know about mitigation the more they trust it, but instead demand to know more about *how* people make sense of their "miti-gated" position and situation. Furthermore, a common understanding among activists and regulators in vapor intrusion debates is that the level of indoor air contamination associated with a decision to install mitigation systems has varied among sites and sometimes even at the same site, which in turn often raises concerns among the public regarding the safety and fairness of mitigation decisions. If tenant notification laws are in fact an option to attain greater fairness, which I think they are, we need to better understand how and who is ultimately responsible for sharing knowledge about the risks of vapor intrusion, mitigation, and even the law of tenant notification itself at the local level. Public meetings and flyers can work, but according to my survey results, renters in the plume do not seem to be showing up to public meetings on the IBM-Endicott pollution issue. In fact, I found that while 37 out of 50 homeowners surveyed (74%) had attended at least one public meeting, only 5 out of 31 renters (16%) surveyed had ever been to a public meeting. I suspect that perhaps one of the major reasons for a low turnout among renters is in part due to the fact that renters have not "invested" in the plume, as homeowners have, and so keeping up to speed on the progress of the remediation effort—the usual focus of the public meetings—is of little interest. Again, this is just speculation, but it is clear to me at this point in the history of the IBM-Endicott debacle that we need to better understand the differences between renters and owners in order to better grasp the unfolding political ecology of risk and mitigation. Of course, another dimension of this political-ecological diagnosis is the emergence of local grassroots action to better "protect" Endicott's plume residents, which is the focus of the next chapter.

6

Grassroots Action and Conflicted Environmental Justice

I just want people to know. People need to know that there is a problem here. People need to know that it ain't gonna be covered up.
—Plume resident and community advocate

Advocates are . . . situated in reciprocal relation to other advocates, even if geographically distant, whose intended as well as unintended actions influence what is perceived as good and possible.
—Fortun (2001:16–17)

The IBM closure in 2002 and the increased public disclosure of IBM's toxic legacy, especially TCE and the threat of vapor intrusion, led to the emergence of several local advocacy groups. This grassroots advocacy developed in the same way that other environmental health and anti-toxics movements (ATMs) have emerged: knowledge of toxic contamination enters into community health etiologies and becomes the grounds for community concern and action. First, plume residents Bernadette Patrick and Sharon Oxx formed a group called the Citizens Acting to Restore Endicott's Environment (CARE) to petition legislators to help plume residents. The IBM contamination led Patrick and Oxx to rethink their own experience with the ambiguity of local health issues. Patrick's daughter developed Hodgkin's disease as a teenager and Oxx's daughter developed bone cancer as a teenager, and their advocacy emerged in response to wanting answers about the cause of their daughters' health problems.

At the same time that CARE was organizing, residents Alan Turnbull and Edward Blaine together developed the Residents Action Group of Endicott (RAGE). Allan Turnbull decided to become an activist after learning that his wife Donna, a non-smoker, developed throat cancer that led to the removal of her salivary glands. Upon learning of the IBM plume and his wife's health condition, in his words, "I decided I was going to get to the bottom of this. The more questions I asked, the more questions I had, and the more questions other people had" (Grossman 2006:100). RAGE went on to develop a website, prod the state and IBM to cough up information and augment their efforts to address residents' concerns, and even received an EPA award for their efforts in community organizing in 2005. Ed Blaine, RAGE's other founder, helped file a local class action in 2008. He is the first listed plaintiff on the pending class action, which consists of more than 1,000 plume residents represented by seven attorneys. Much of RAGE's success and ability to connect with political officials came from its former member, Donna Lupardo, who became a New York State assemblywoman in 2007.

The CARE and RAGE organizations are no longer active. Today, the local organization supporting the concerns of plume residents is the Western Broome Environmental Stakeholders Coalition (WBESC). Filling the shoes of RAGE in particular, this coalition developed in 2006, and is made up of about a dozen members. The primary goal of the WBESC is to help organize public meetings and host guest speakers (e.g., agency representatives, scientists, and activists). The chair of the WBESC told me in an interview that one of the main reasons the WBESC developed was to "make it easier for the agencies to work with the community, because CARE and RAGE were both requesting the same information and each demanding the attention of the state agencies, so we decided it was better to combine them, and that is how we got the coalition." Wanda also explained to me that Turnbull had some health problems, including a hip replacement that made it a struggle for him to attend meetings, resulting in her taking over as Chair and lead contact person for the WBESC.

While public health has always been a concern for residents impacted by the IBM plume and for local activists, the WBESC has focused much of its recent efforts on the occupational health of former IBMers who worked at the Endicott plant. With the help and expertise of Richard Clapp, who has done groundbreaking epidemiological research on occupational health issues among IBM workers (see Clapp 2006, 2008), the WBESC has worked hard to get a study by the National Institute for Occupational Safety and Health investigating over 25,000 corporate mortality files for workers from the former IBM plant. NIOSH did a feasibility study when residents first raised this concern, and after completing the study determined it was feasible to do the study and that it would cost an estimated $3.1 million. Residents and activists are still waiting for the results of the in-depth study, which was estimated to take up to five years, depending on funding and the "cooperation" of IBM. The most recent epidemiological report (Clapp 2008) focusing on cancer mortality for IBM Endicott plant workers who worked at the plant between 1969 and 2001, found significantly increased mortality due to melanoma and lymphoma in males and modestly increased mortality due to kidney cancer and brain cancer in males and breast cancer in females.

The most recent effort of the WBESC has been to get the NYSDEC to further investigate vapor intrusion at the current Endicott Interconnect (EIT) facility. One WBESC member and union organizer with Alliance@ IBM continues to prod the state to further investigate present exposure issues for EIT workers. This WBESC member argues, as do many activists I interviewed, that it is absurd that EIT is not required to install the same vapor intrusion mitigation systems found in the IBM plume area. Many scratch their heads at this fact. So, despite the tremendous success of local advocacy groups, the IBM contamination and the struggle of plume residents persists, even while TCE vapor intrusion "risk" has been determined, according to state agencies, a "mitigated" problem.

As Edelstein (2004) contends, and I think rightly so, contamination is both disabling and enabling. The IBM contamination in Endicott has resulted in many *disabling* consequences (e.g., health issues with uncertain causes, stigmatization, community corrosion, hopelessness, economic hardship, stress, etc.) that may never be mitigated or remediated, despite the ongoing efforts of IBM and the NYSDEC. Amid this toxics rupture, people have taken action—some vocally, many more in private—resulting in a citizen response that has *enabled* a "sense of community" or at least some sense of collective intentionality. In many ways, citizen action in contaminated post-IBM Endicott has reinforced the idea that people "do not live in secluded cocoons of [their] own" (Sen 2009:130), that when a community-level problem arises and a sense of duty emerges, citizens decide to become actively engaged in a "politics of enunciation . . . [whereby different] . . . identities come together" (Allen 2003:21) to share their stories of struggle. Furthermore, I think it is fair to say that the advocacy in Endicott is more than anything a "response to temporally specific paradox" than it is a "matter of shared values, interests, or even culture" (Fortun 2001:11). In other words, the locally situated citizen response to IBM "productivity and plunder"[1] in Endicott has its own complexities and paradoxes, its own fomentation process, as David Harvey would put it.[2]

When I returned to Endicott in the summer of 2009, I expected plume residents to make up the majority of local activists engaged in the Western Broome Environmental Stakeholders Coalition (WBESC). Instead, what I found was that only one of the activists listed as a member of the WBESC actually lived in "the plume" and with the durable ambiguities that come with living in a "risk" zone and in a mitigated home. I was told by one activist that this was not the case when activism in Endicott was under the leadership of the Residents Action Group of Endicott (RAGE) and Citizens Acting to Restore Endicott's Environment (CARE). Even though RAGE's founders lived and still live several blocks west and outside of the "official" plume zone, plume residents made up the majority the citizens involved with CARE and RAGE when they began to surface in 2003 and 2004. So, what I came to realize and what I hope to expose in this chapter is that augmenting the "politics of enunciation" (Allen 2003:21)[3] so commonly found in contaminated communities, is the tension between local activists—the people who "make it their duty to speak for the people" (Bourdieu 1990:186)[4]—and the "community stakeholder" model informing contemporary environmental politics. Fortun (2001:10–11) helps explain this point of friction:

> Who inhabits this world? How, once social and territorial boundaries have been destabilized, can one discern what a community is and who is part of it? How does one account for the way disaster *creates* community? Within environmental politics, these questions are usually responded to in terms of "stakeholder"—groups of people who have a stake in decisions to be made by corporations, government agencies, or other organizational bodies within which decisions by a few can affect many. A stakeholder model recognizes different social positions and different ways of

perceiving both problems and solutions. The model has merits. It has pushed corporations to include nonshareholders in their calculations of stakes. Workers, people living near production facilities, vendors, and others have been offered a place at the table. It has also helped draw once marginalized players into policy formulation and evaluation. . . . But there are problems with the stakeholders model [that the IBM pollution conflict in Endicott] makes visible. The value of the model is its recognition of difference. In use, however, the goal is usually to manage difference by forcing diversity into consensus. And each stakeholder community is usually considered to be epistemologically homogenous and epistemologically consistent. Members of any given stakeholder community are assumed to think alike. . . . Like most pluralist models, the stakeholder model can't seem to tolerate much complexity—or much dissent.

With this critical "stakeholder" perspective in mind, this chapter draws on activists' narratives and lived experience to better understand the subjectivity or point of view of active community stakeholders in Endicott. With a general focus on the intersections of subjectivity and intentionality, this chapter, then, aims to "make sense" of these citizens' advocacy efforts, and draws on narratives generated from in-depth ethnographic interviews with local activists. Several investigative journalists (Gramza 2009; Grossman 2006) have helped popularize and tell the story of environmental activism in Endicott. This chapter aims to deepen that understanding by drawing on extensive narratives from in-depth ethnographic interviews to highlight activists' perspectives and to illustrate the ways in which this local activism converges with "boundary movement" theory and bears witness to the micropolitics of environmental justice alliance building that expose the messy "intra-community tensions" (Horowitz 2008) that challenge the spread of environmental justice politics.

Grassroots Action: Narratives of Duty and Dissent

In my interviews with members of the WBESC,[5] I asked general questions used in ethnographic studies of social movement, advocacy, and activism: Why did you get involved in advocacy? Why did you *become* an activist? What has becoming an activist been like? How do you see your efforts changing or not changing the current and future situation? These questions aim to intensify our understanding of not only the acting subject, but also the "intentionality" of human agency or "the ways in which action is cognitively and emotionally pointed *toward* some purpose" (Ortner 2006:134; emphasis in original). Most activists I interviewed told me that they became involved to help pressure IBM to do a better job of cleaning up the plume and pay greater attention to the occupational health exposure issues for workers who used to work at the plant. The latter became a quick focus because many of the activists I interviewed who had started with the Residents Action Group of Endicott (RAGE) and were now WBESC members were also former IBMers.[6] What follows is a close-up look at some activists' narratives I collected during the 2008–2009 field season.

Inspired by critical ethnographies of advocacy and environmental justice (Allen 2003; Checker 2005; Fortun 2001), I wish to provide a narrative sketch of local activists that attends to intersubjectivity,[7] the micropolitical, and therefore to various points of convergence and divergence between advocates. I intend these narratives to "function as both story and rhetoric, identifying the various stakeholders as part of an interpretive community" (Allen 2003:22). As stated in the introduction of this book, I think of Endicott as a post-IBM community perplexed by the uncertainties of the local environmental and occupation health risks of TCE exposure and the ambiguous efficacy of mitigation and remediation decisions and technologies. Activists engaged in this ambiguous battle take on a certain interpretive role, and their narratives help make the IBM contamination conflict intelligible. These narratives also showcase the way in which advocates are certain about what is "good and possible" (Fortun 2001:16–17), even if their interpretations remain confused or their actions do more to reinforce or stabilize uncertainty than to deliver the Truth, the certainty of matters.

My hope is that this chapter provides a discursive space that anchors the debate in the narratives of those engaged in actively making the debate attentive to "community" concerns. Furthermore, these narratives do more than provide accounts of activists' lives; they expose connections between activists, between their converging and diverging concerns, interests, and virtues. They showcase points of transparency, agreement, and corresponding intentionality. But, as discussed in previous chapters focusing on risk and mitigation experience, residents' narratives also help intensify our understanding of the place of difference (or what a difference difference makes) in the social world. "Like 'being' according to

Aristotle, the social world can be uttered and constructed in different ways: it can be practically perceived, uttered, constructed, in accordance with different principles of vision and division" (Bourdieu 1991:232). What follows are examples of how activists utter and construct their advocacy identity and what it is like for them to be engaged in activism and advocacy work in Endicott. I discuss activists who have been most vocal and involved in the IBM-Endicott contamination debate and have selected narratives highlighting points of convergence and difference between this small network of activists who, in many ways, exemplify actors engaged in advocacy that blurs the boundaries between anti-toxics and environmental health movements and movements for environmental justice.

Shelley's Story: The Outspoken Republican and Depoliticizer

Shelley was the first WBESC member I interviewed when I returned to Endicott for fieldwork in the summer of 2008. Reflecting on her time as a nurse for IBM's first aid unit between 1969 and 1979, Shelley told me she had firsthand experience with IBM workers and chemical exposure issues: "Wherever they needed me I would go. We had about 23 on staff and 3 full shift . . . I was the person workers would go to if there was a body spill. They had to come to my first aid. They would go directly into the shower and they would be cursing up a storm. Some of them burned and many were released for blood work." In addition to being a nurse at the plant, Shelley was a Broome county legislator for 20 years, and has been described by local journalists as an "outspoken Republican." Sitting in her living room on Endicott's hilly "North side," located outside of the IBM Endicott plume, she explained why she got involved with RAGE and told me about the history of the emergence of the WBESC:

> You know, I may not live in the plume, but I know what those people are going through and the anxiety level, I am sure, with the landlords and home owners. You know it is probably not something they are thinking about all the time, but the next time you get a little pain in your gut and you go to the doctor, it is in the back of your head, "I wonder. I wonder." But, I refuse to have this thing politicized. I don't know how many times I have to remind a few of our members that this is not a democratic issue. And Alan Turnbull, the guy who got the whole RAGE thing started, he's

a fantastic guy. He is the guy who got me involved and he said we really need you. I attended a couple of meetings and I realized that there were several constituency groups, from Sierra Club to the CARE organization to RAGE, and a whole lot of people who weren't organized. Well, I also know that our resources are limited here in New York. It doesn't make a whole lot of sense to have officials responding to several constituency groups every single month. We are important, but we are not that important, but together we can say to them, "Instead of coming down six times a month to Broome County, why don't you come down once and all of our groups could be part of this coalition." They liked the idea. I did not know that ATSDR, NIOSH, and the CDC already had a component ready as far as community groups. They came down with a strategic planning group, but you know a strategic planning group doesn't mean anything to the constituent. I said, "You know, no one knows what that is. A strategic planning group for what?" I said "We have to have a name that identifies us." Strategic planning group sounded kind of stupid. So, that is where Western Broome Environmental Stakeholders Coalition came from. Everybody is using the word stakeholder nowadays. It is all over. You have a stakeholder this, a stakeholder that. We needed to identify ourselves and we tried to stay western Broome oriented because this is where the majority of the population and where most of the taxes are paid from.

Finding myself still contemplating her depoliticization comment (I refuse to have this thing politicized"), and reminded of the "anti-politics" rhetoric of neoliberalism (Büscher 2010; Ferguson 1994), I moved on to my interview questions, even though I found this potent "political" statement interesting and a good example of the "micropolitical ecology" (Horowitz 2008; Little 2012b) of environmental advocacy. Moving on, I asked her, "So, what got you interested in the IBM pollution issue in particular? What sparked your interest in advocacy?" She replied:

I did 40 years community service while working and raising my family. I worked in almost every single community group you can think of. But, cancer to me was the thing. When I was a student nurse, back then in 1963 in the pediatric unit, babies didn't come out of there. Leukemia was leukemia. Deadly leukemia. I saw babies with colon cancer and babies with brain cancer. Part of the training was autopsies. You are standing there and you are seeing this horrible disease ravaging people, affecting little ones that didn't do anything. It always stayed with me. You know we had young people that we took care of and you know there is an odor of cancer that you can't get out of your mind. With IBM, even though I did first aid work, they asked me because I had free time to take care of some IBMers that had terminal cancer. I loved nursing. You know what, I didn't mind doing terminal care nursing. Now we call it hospice. IBM liked the fact that I knew IBM, that I could talk to the employees about IBM. Also, I have kind of a wacky personality. For example, if I was working there on Halloween I would come in with a mask. They loved that.

Like other WBESC members, Shelley knew people who worked for IBM that had died of cancer. Like other activists, she experienced things that "always stayed" with her. For example, she told me about her neighbor and friend who recently passed away, and I sensed that her close relationship with him fueled her advocacy and especially inspired her to prod the National Institute of Occupational Safety and Health (NIOSH) to do an occupational health study of IBM workers who had worked at the Endicott plant:

Seeing the cancer and knowing it . . . I have a strong faith, believing there is something bigger than us out there, and I am always questioning . . . I just lost my neighbor and he worked for IBM for 27 years in maintenance. Here I have a little prayer box with his picture in it. [She showed me the picture which was posted in the obituary section of the local paper.] He was much more handsome than the picture shows. He got cancer about five or six years ago. We thought he had it licked. He was my buddy. We would always exchange things and talk politics. . . . We were always talking back and forth and if there was an issue that would come up politically we would chat. He had a good sense of how I would go with the vote. He kind of typifies the Broome County person [By this she meant he votes Republican]. I miss him terribly and he never really even called me Wanda. It was always "Hi neighbor." But, when he talked to my husband he would say "How is Wanda?" He didn't have much property to sit out on and relax, but he loved to be outside. He would put out a little chair on the side of the property and we used to sit and talk. As he got sicker, he didn't come out, you know. He couldn't make it around the corner, so he started to sit right outside the garage. He would just sit there.

Well, this man saw the [feasibility] study that they [NIOSH] had initially done with all the chemicals used at IBM. When I saw the chemicals and the cancer they reported I was just sick. He asked to see the NIOSH document. . . . He never complained about a situation. Whatever it was he did not want to see the resolution or anything like that. So, I gave him the study and about a week later he comes on down and handed it to me and looks me right in the eye. He looked me right in the eye. He sat down on my deck, which was very unusual, and he said, and these were his exact words: "I worked with every single one of these chemicals for 27 years. You can't stop what you are doing." He didn't say "Do this because I have got the cancer" or "I'm dying." He just said, "You can't stop what you're doing." You wanna know what? I'm not going to stop.

We have got the NIOSH study ready to go under our belt. We are going to be helping with the protocol. We have studied almost everything we can study here and that is about as frank as I am going to get about it. There is no more issue to study down here. We have got it. We know, now we have got to prove it and we want that study of IBM workers.

Despite her commitment to the NIOSH study and making sure the WBESC continues to hold meetings and provide information for the community, Shelley also shared with me her sense of "burnout," hopelessness, and her feelings about the durable ambiguity of the future of the WBESC. She was especially vocal about struggling with keeping the WBESC afloat and even losing the personal motivation to "keep going" as the WBESC's leader:

SHELLEY: In about another five years I will be 70. It sounds crazy to say. I hope this work keeps me healthy. I had some close calls here with back problems, you know. I told the group about two months ago, "You know, it is getting to the point where sometimes I sit and say, what the hell are we doing here? What is the use?" I was very frank with them. I said. "I have no budget. I can't do half the things I would like to do because I have no money." Nobody comes forward and gives me money like they do at church. Now I just tell the officials, "You got a report for us, you better have it for everybody. Don't just

give it to me. Everybody. Because, I am not going to print the thing
out." I don't know anyone that wants this job.

PETER: So, what keeps you going?

SHELLEY: People keep telling me quietly, "Keep going, keep going. No one
else is going to do it." And I know that and that is how it was for
Alan Turnbull. This man delayed hip surgery for five years to keep
RAGE going. He could hardly walk by the end of last year. You gotta
understand that all of a sudden your life starts to consume you. You
start saying to yourself, "Gee, I better not have that hip replacement
because it is going to be about three months of recovery and who
is going to handle the meetings? Who is going to make sure this is
going on and that is going on?" That is how we are. I was flat on my
back for 2–3 months at the beginning of the year and I told them
right out either we are not going to have a meeting or if we have a
meeting you are going to behave yourselves, you know. It is almost
like I am a grandma here.

Shelley is not only open about being Republican, but she also openly
stresses that as the leader of the WBESC, she needs to control for the
threat of the politicization of the IBM contamination conflict. For example, when she can't make meetings, she finds herself worrying about the
so-called political contamination of the IBM pollution debate. Again,
while she made clear that her mission was to depoliticize the issue and
make the WBESC apolitical, her sustained defense of the need for an
apolitical WBESC had its own micro-political tone:

It's like if I can't be there, I better not hear of anything going wrong. I tell
them to get back to me. That is how it has got to be. My secretary thinks I am
a little too harsh on people. But, when you are out there and you say something political in that meeting room, I am going to get back at you. Don't
go telling me that only the Democrats care about the environment. That is
not true. We all care. We have one guy who is a democratic activist and for a
while I had to tell him that the WBESC is not politically oriented. If someone wants to go ahead on their own with something that is one thing, but
this is a coalition. In order to do something with the name WBESC on it
you have to have the agreement of every single participant. . . . That's how
it has gotta be and I learned that when I was a Broome County legislator.

> There were many things that I thought were wonderful, but I couldn't say
> that Broome County approved of it because I was only one in 19 legislators.

While the depoliticization of the WBESC was a strong theme in my interview with Shelley, other members of the WBESC believe politics figure into the debates, as we will see later on. But, to be fair, it was my sense that for all the activists I interviewed there is a strong common grounding, something of a collective concern, that dwarfs the inevitable micropolitics of the WBESC: TCE contamination is a problem, the WBESC will make sure IBM, the NYSDEC, and NIOSH will do their job and "stay on top of the issue," as Shelley put it. As Nicholas Freudenberg wrote in his classic book *Not in Our Backyards!* "If activists understand why they need to work together, they can find the energy to address the issues that divide them" (Freudenberg 1984:238). This seems to be a welcome possibility for contemporary activists in Endicott, even amid real differences between coalition members.

Frank's Story: Refusing to Be a Silent IBMer

At 59 and a resident of Endicott since 1966, Frank spends long hours at the Alliance@IBM[8] office even though he actually works part-time. He describes himself as a dedicated union organizer and active volunteer, in addition to being a web tech, and a professional musician. "I've been a Rock and Roll singer for 44 years. I was a member of the Human Beings. You are too young to remember, but we had a hit record in 1968 called 'Nobody But Me.' It was just a one hit wonder." He worked at IBM as an electrical engineering technician for just over 28 years, starting in 1974. Like Shelley, Frank does not live in the plume. As he put it, "I'm not affected directly by the plume. I may be indirectly if it is in the water system, because I get Endicott water, but based on all the meetings, apparently all the water is testing to be ok. I am not sure all of the good news they tell ya, but outside of that, no I am not affected. I am not ventilated."

When I was in the process of corresponding with Frank to set up our interview, he asked that I send him my National Science Foundation (NSF) proposal to get a better idea of what I was up to. He requested this, because, as he put it, "A lot of people have come here to do a study, and we don't know what they are really doing. They leave and we have no idea what they are doing." When it came time for the interview, the first thing he said was, "Well, I read it and you've got it pretty much covered." I said, "Well, that's funny, because I feel as though I haven't even started because I haven't been able to talk directly to activists like you about this issue." He responded by explaining that his intentions as an activist are rooted in his concern for the health of his fellow workers at the old IBM Endicott plant:

> Well, my connection as a stakeholder is with the workers and the former IBM workers at the plant. That is where my connection is because I was exposed to a soup of different chemicals. As a tangent note I can't say I have suffered directly from it, but I know because I was in manufacturing for some time that people like me who worked in manufacturing were not so lucky. So that is where my connection is.

For Frank, speaking up and being involved with Alliance@IBM, RAGE, and WBESC has created "personal" problems. For example, he blames himself for letting his activism get between him and his wife: "I mean on a personal note, my wife and I still have discussions that turn heated about my continuing involvement as a community activist and as a union activist. I mean, it has put a huge strain on my family and I take the blame for that." His wife supports him, but he knows how IBM and EIT feel about activists like him: "I think IBM and EIT and anybody that hates unions and vocal activists like myself would wish that I would just shut the hell up, you know."

Frank played a critical role in getting the current NIOSH study, despite the obvious role of congressman Hinchey. He "helped to get pressure to get it done," but contends that the person really responsible for prodding NIOSH to do the worker health study was congressman Maurice Hinchey. "The people that he knew in the government was critical to getting this thing going. Hinchey believed whole heartedly that IBM workers and not just the residents, but the IBM workers had suffered greatly and that it was not well known, period." Frank explained that workers have serious reasons for not sharing or talking about "what they know," but this doesn't stop him from trying to get workers to speak up. Furthermore, he explained that "it gets tough being an activist" when former IBMers end up at meetings "pointing their fingers" and telling "multiple anecdotal stories":

The perception among people here is that there are these agencies and companies to fear so they are not gonna open their mouth and they are not gonna step forward.

Notwithstanding, many of those people are retired IBMers on a pension and fear they are going to lose their pension or IBM is going to take that away from them if they open their mouth. I mean people have said that to me directly. They say "I don't want to say anything. I am drawing a good pension. I am ok and I'm not sick." They have said to my face, "I think your right, but I'm not gonna tell you what I know." I say "You're not going to tell me what you know, oh my God." This is what we have here. . . . I don't even have half of the information that other people have that could put these guys in jail, potentially. Potentially. But I know the people that know this information and sooner or later I am hoping I can get them to talk about it. I don't know that IBM or the agencies know who these people are. Some of them are vocal. Some of them are angry and open about how they feel. Some of them come to the WBESC meeting and they point fingers at the New York state agency people and accuse them of all kinds of atrocities and get into a shouting match. You know, I am glad they are bringing the information, but this is where it gets tough being an activist and being on the board and having a balance, because like one person that I can think of has showed up to the meetings pointing fingers and proceeded to tell multiple anecdotal stories about what IBM's trucking contractors and toxic chemical disposal contractors did back in the 1960s and 1970s. I mean he basically tongue-lashed the NYSDEC guy and tried forcing him to send people into the hills of Endwell which is up near the IBM Homestead and where the two golf courses were.

Frank went on to explain more about the cohort of silent IBMers whom he wants to hear speak up. His narrative highlights the extra complications that the internal politics among IBMers brings to the issue, even though he "hates" to bring up the politics at play in the debate:

I have spoken to people that work in offices on the third or fourth floors of some of those buildings at the IBM Endicott facility who believe that IBM has basically tried to cover all this up or make it right to make sure their image did not get damaged for what they did, but they are still unaffected by the chemicals because they worked in the office. So they are not really exposed. However, maybe the person in the cube next to them has the opposite attitude, where they think "Oh, IBM is not to blame for this. Some of these laws for some of these chemicals were not even enacted back in the 1950s and the 1960s. The government is to blame for this." You know, and I really hate to say this, but my perception is that it starts becoming political. When you hear both sides of two people, for example, and I have, that work in the same office and neither of them are exposed to the chemicals. . . . it's amazing that each sees the opposite view. And it all kinda boils down to left and right, or Democrat or Republican, or liberal and conservative. I hate to say that, but that is my perception. But, really when you listen to their arguments, and some of the arguments are not their own words, some of them are from people that they talked to, and some of it is their own assessment. It is just the politics and it isn't glaring. It isn't glaring. I mean I have listened to a lot of people talk about this issue and I know I am biased about it on several levels, but I try to be objective and listen to the point their [sic] making. But every time they make a point there is a taint of politics in their discussion. I guess I can say from my own little analysis of it, it is that if it is somebody who is against big government, you know, a Reagan era kind of person, they tend to blame a lot of this on too much government interference in IBM's affairs at the bottom line. They say they are trying to make laws retroactive that weren't formulated during the time that the spills all took place. Whereas on the other side, on the left if you will, it is quite the opposite. These people end up making statements or imply that they think that business is evil, you know, ultimately because their goals are narrow and focused and they don't take in the human factor and even consider the human factor a thing. They don't consider it a thing. That also reflects their attitude about their work life, you know, how IBM treats them and its workers.

Frank went on explaining how he feels about the milieu of politics, even though he felt his sample size was probably too small to really know, to have an "objective view":

It seems to me, and my sample size is not documented and probably not big enough, but it seems to me that the people that I talk to that were happy with their job at IBM had not been exposed to any chemicals or toxic materials in the workplace. They are happy with their salary and what IBM is doing as far as a company, and they well just say that it's the government's fault. They tend to be republicans or conservatives. I think that's it. I know that's really not an objective view. And it is the same on the other side. Like I said, I am admittedly biased and I kinda veer to the left, but there are things that IBM has done that I don't think other companies would have done, so I have to applaud them for that. That is great.

My sense is that while Frank self-identifies as a leftist union organizer, he is more modest and "cooperative" when it comes to how he talks about IBM and the NYSDEC vis-à-vis their involvement in the cleanup effort. He still thinks the worker exposure issue needs more attention, and he, like other WBESC members I interviewed, is waiting for the final NIOSH report, which will be completed as early as September 2014. Despite moments of applause for the work IBM and the responding government agencies (e.g., NYSDEC, NYSDOH, ATSDR, and NIOSH) are doing, and the attention Endicott is getting when compared to other communities affected by TCE vapor intrusion, Frank remains perplexed. He has attended all the meetings, and has interacted with all the core government officials working on Endicott's Superfund site. His interactions with these have been, at times, confusing and have left him wondering where things are headed:

> There have been times when they have met with WBESC or RAGE and they have been resistive to doing things and have blamed it on a lack of resources or lack of sample size or anecdotal data being irrelevant and unscientific. I mean there have been several times where I have been in a direct conversation with either the NYSDOH or NYSDEC or ATSDR and I have come away from that conversation saying, "What just happened here? Where are we going?"

John's Story: Being the "Squeaky Wheel" and Information Disseminator

John's activism derived from a mixture of dissent, the excitement of "embarrassing" IBM, and a personal interest in "making a difference" and testing to see if activism could make a difference:

The reason I wanted to get involved over here was because I wanted to see if one person can make a difference as an activist. Could I make an impact? The other thing is I wanted IBM to pay for the crime they committed. I wanted to use them as an example to show them that they can't get away with that. We are trying to publicly embarrass them and trying to make them clean it up.

John, aged 56, worked at IBM as a computer specialist for 20 years, and feels he lost his job at IBM because he was "considered a safety advocate at the plant and the big boss did not approve of that." He has lived in Endicott for 45 years and now lives in a house "just on the border of the plume, although the house has never been tested. They tested around it, but they estimated that it was not a problem here. Less than a block away people have it [a vapor mitigation system]. I joined the class action suit because no one really knows what the contamination is like here because they didn't test it."

John is one of three WBESC members I interviewed who is engaged in what Shelley and more modest WBESC members would call, for lack of a better term, "radical" or "leftist" activism. Embarrassing IBM has been the primary intention of his activism. From the beginning, he has aimed to be a "squeaky wheel":

Before RAGE was even formed, I wrote a letter to the editor and they used a pen name. Like I said, I did not want to get in trouble, so I used a pen name. I wrote my letter, saying IBM should be responsible for cleaning up these chemicals and the newspaper decided not to print because they thought it would be a liability. In other words, IBM could sue them. So I kept on them and said you need to print this letter because of the public health concerns for the community. They did. I finally got them to print it. I think that helped get other residents involved. I mean they got involved on their own as well, but I think my letter helped draw some attention. It helped put some political pressure on IBM. After that I would get on the local radio. I would send out emails to politicians to get involved. Then one of the politicians had a meeting at the visitors' center and that is where Alan Turnbull got community members together to form RAGE. I was at that meeting. I got on their distribution list, so I got newsletters and got involved with them a little bit. I was still working at IBM.

I was talking to an activist once and he said, "The way to make this issue heard is to be a squeaky wheel and make a lot of noise." So, I created a distribution list of thousands of newspapers and reporters and editors around the country and when I heard there was going to be an article in the newspaper, I would send it out to thousands of people, including politicians. I spent months, years, just building this distribution list and I would write my own letters of concern about the residents in the contaminated area and also information and concerns about chemical exposures in the workplace because what I experienced in the workplace working on some of these chemical machines that expose workers to many different chemicals like copper chloride, ethylene chloride and a lot of these other nasty toxic stuff. Initially, before the attention was on the chemical spill which originally occurred in 1979, I began to form a relationship with reporters at the local press here. If I would hear things in the workplace, I would express my concerns to them. Sometimes we would be running these machines and IBM would not keep up on the machines' maintenance because IBM wanted to get their product out and get their quota and the machines would have leaking seals. The chemicals would spill out in the mote underneath the machine. You would be breathing in these fumes. I would call OSHA [Occupational Safety and Health Administration] and they would always come in wearing protective suits and slippers and say "Oh, it is within OSHA standards." So basically, they wouldn't do anything about it. Well people with NIOSH at the time said that the OSHA standards are outdated. It made me realize that not all the companies follow the standards because you have a government organization telling them they are within OSHA standards and the corporation thinks they are safe, they are complying with OSHA standards. So, I became not only an activist against IBM and the chemical spill but against OSHA and their standards.

Mike's Story: Listening with an "Open Mind"

Mike, a retired engineer, used to work in the field of power plant water chemistry. Even though he is 57 and retired, he still does contract work, and is a landlord with property in the plume. He has lived in Broome County all his life. He grew up in Endicott, where his dad worked for the Endicott-Johnson Shoe Company. His parents migrated to the United States from Italy (south of Rome). Mike lived in Endicott until he was

27, when he got married and moved to Hillcrest,[9] about ten miles east of Endicott. As an engineer, Mike tells me that "Most of the time, I understand what the engineers and scientists are saying at the meetings." He went to Broome County Community College, which was Broome Tech at the time, and then went to Clarkson University in Potsdam, New York.[10] Mike's mom worked for IBM for about six months during the 1940s. He does not live in the plume now, but he did grow up in what is now the area impacted by IBM's TCE plume. Mike shared with me his reason for getting involved with the WBESC and why he decided to become a stakeholder:

> I grew up in downtown Endicott. My mom and dad had property there and I lived there. So I became interested for two reasons. One, obviously the value of the property associated with the plume, and two, the health concerns. My mom lived on Madison Ave and my sister lives right next door. So I wanted to make sure things were going ok for them. I got involved with the WBESC right when it started. My role is secretary. I think we were joking the other night about not getting the minutes out. But, I retired in May of 08 and have part-time work, and so now it feels like I am busier now than when I was working.

Mike informed me that one of the strengths of the WBESC is that it is an advocacy group with a wide vision, and a group that has been successful in reviewing and making sense of "complicated and technical issues": "I think we have done a lot to make sure a whole range of different aspects have been covered. I think we have done a good job with the complicated and technical issues. We have done the local health studies and we have been working with the health department on that. I think we have done a really good job reviewing things." He went on to explain that

> in terms of looking into the future, we want to continue pushing for this workers' health study because it looks like the area itself may not have been affected by the plume as far as health concerns. I say "may" because it is really difficult to pinpoint whether it is or not. But, with the workers' health concerns, there seems to be something going on there. I think we should push that. The other aspect is the remediation. Right now we don't have another process for remediating, other than the flushing

process that they are doing. That's improving, but it won't, I don't think, come to an absolute zero for TCE in the plume. So, that would be the next thing, to push for finding better remediation.

Mike made sure I understood that he "doesn't normally get involved with advocacy organizations." So, to better understand what the experience has been like for him as a "new" activist, I asked him, "What has that experience been like and has it changed anything for you?" He responded:

Well I am open to other people's opinions and that's really what the whole thing is about. So I try to listen with an open mind and what we try to do, and what I try to do, is try to keep people focused on what the issue is. Because a lot of the time at these meetings, and within the group in general, people start to bring up other things. Some people, for example, want to get involved with the drilling, the natural gas drilling. Well, it's like this group is not for gas drilling. Yeah, gas drilling environmental aspects are very important, but let's keep focused. That's what I have seen. When we first started, we didn't know exactly where we were headed. So after we heard of a number of different reports we started to go through a number of iterations about where we should be and after all of that it looked like the most important thing right now is the workers' study. The reason was because we heard about a number of different studies that brought up these concerns. It looked like the environmental aspects of the plume were being taken care of but the workers' concerns were not. So after hearing various speakers mention this concern, we figured that this was where we can push the agencies a little bit more to find out, to do a better investigation of those concerns.

Even though the decision was made by WBESC members to concentrate their efforts on the NIOSH study, Mike points out that at the beginning he and other members of the WBESC didn't know exactly where they were headed. Of course, ambiguity can be a common struggle for coalition-building. But, this directional confusion was not the case for some environmental advocates engaged in early pollution activism in Endicott. For one former RAGE advocate, her advocacy was motivated by not only personal but also political interests.

Diane's Story: Juggling Environmentalism and a Political Career

Currently representing the 126th District of the New York State Assembly, Diane grew up in Endwell, where she still lives. While many local activists I interviewed said that Diane won her state assemblywoman ticket because she was involved with the IBM contamination debate, she explained to me that her concern with the safety of her own drinking water figured strongly in her decision to get involved with local advocacy groups:

> I have been an environmentalist for most of my adult life and when I found out that the home I was living in Endwell, NY got its water from the Village of Endicott water supply I started meeting with advocacy groups, which were at the time CARE and RAGE. I came to these groups with my organizing assistance and my knowledge of environmental issues. When I joined these advocacy groups I was a county legislator and I was given the assignment basically as legislative director of RAGE and policy director of RAGE. I was at the time contemplating running for NY State Assembly and I was able to focus our direction based on my experience. So that was my initial involvement with RAGE and then that sort of morphed into the stakeholders planning group. This is the WBESC. Before joining the coalition I was elected to my first term in the NY State Assembly and so I had to step down because there would have been a conflict of interest. I became a different kind of stakeholder, a stakeholder with State assembly representation. So I had stepped down from all board director positions I had belonged to, including RAGE and others.
>
> Obviously RAGE has experienced a lot of success. We were able to get congressmen Maurice Hinchey to reopen that site which was previously closed. We were one of the first sites in the country to benefit from the development of vapor intrusion, with many homes and businesses in the community outfitted with vapor mitigation systems over the 300-acre site. The RAGE group continues to provide guidance to the community as they go through this. RAGE is quiet now because the initial goal has been satisfied and I think it is probably important for them to renew their focus around how people are in fact coping with the ongoing remediation, but that has not been something that they have thought through.

Among the successes of RAGE, which is now the WBESC, Diane lists "the development of vapor intrusion." I heard this rhetoric of vapor intrusion progress in my interviews with other government officials and activists engaged in vapor intrusion debates.

In my interviews with NYSDEC officials, scientists, and regulators with the Interstate for Technology and Regulatory Council's Vapor Intrusion Team, and EPA regulators, they all tend to agree that the work and research on the IBM-Endicott plume has helped advance "scientific" understandings of the vapor intrusion pathway. For the plume resident and activist I discuss next, it is difficult to think that plume residents have witnessed "development" or progress when one is stuck living on a contaminated property, even if that property has been determined mitigated.

Ted's Story: A True Plume Activist and Message Maker

When I returned to Endicott in the summer of 2009, I expected plume residents to make up the majority of local activists engaged in the WBESC. Instead, what I found was that only one of the activists listed as a member of the WBESC actually lived in "the plume" and in a mitigated home. I was told by one activist that this was not the case when activism in Endicott was under the leadership of RAGE and CARE. Even though RAGE's founders themselves live several blocks west of the "official" plume zone, plume residents made up the majority of the citizens involved with CARE and RAGE when they began to surface in 2003 and 2004.

One resident activist and member of the WBESC is not only vocal, but has turned his property into a message board. (I actually helped Ted graffiti a message on the side of his building on Halloween night, 2008.) Ted is 53 and was born in Endicott. Ted, his grandmother, and his father worked at IBM. When I first met Ted at his property across from the IBM facility—where he lives and runs a small business selling coffee and flowers—he showed me an old picture of his grandmother and grandfather with Thomas J. Watson, Sr., that hung on the wall of his coffee shop, which also functions as a smoking zone for EIT workers who pop in from across the street, in addition to the small number of "left over" IBMers still working in IBM's payroll office.

"I'm a third generation IBMer Pete. This is what I know, you know what I mean?" Ted worked at IBM from 1997 until 2001 when he was laid off. "I was only there for a short time, but it sucked getting laid off. I don't care how long you are with a company. Getting laid off sucks." He worked in the compliant pin division, pressing computer components onto circuit boards. He lives across the street from where the 1979 spill occurred and directly above the sewer line that carried the chemicals from the former IBM facility. He bought the building in 2000 and was never told about the pollution at the time of the sale. "I bought a piece of garbage, and nobody told me," was how he described this unfortunate situation. This was the main reason he got involved in local advocacy, in addition to the fact that he had experienced firsthand a violation of NYSDEC law and is a victim of the NYSDEC falling short of its stated mission "To conserve, improve and protect New York's natural resources and environment and to prevent, abate and control water, land and air pollution, in order to enhance the health, safety and welfare of the people of the state and their overall economic and social well-being."[11] When I asked Ted to talk about the advocacy work he has been involved in thus far, he replied:

> TED: Well, I don't wanna say it's a waste of time, but we go to meetings
> and the NYSDEC tells us what they are doing. They get paid doing
> what they do, but we don't get paid. You know what I mean. Right
> now we are trying to get the renters notification bill. We put it in
> with Donna Lupardo [New York State Assemblywoman]. It went up
> to George Pataki [New York's fifty-third governor] and he vetoed
> it. It went to Eliot Spitzer [New York's fifty-fourth governor] and he
> vetoed it. Right now it's on the desk of David Patterson [New York's
> fifty-fifth governor and the state's first African American governor]
> and we don't hear a word about it, ok. Now that's just to tell people
> that you're living on a polluted property. You think these people are
> protecting the people. It's never gonna get passed, come on. So all
> this work and we got one bill passed. One bill that they did pass was
> that if EIT did spill something they had to tell me within 48 hours. I
> think it should be less than 48 hours.
> PETER: What has it been like using your building as a message board with
> all the graffiti and signs on the property?
> TED: The first mayor that was in office when all this came out sold us out to
> IBM. She took money, $2.2 million for the village, who is not even in
> the contaminated area and left all the property owners hanging to do

Figure 6.1
Storefront
Message.
Photo by
Peter C.
Little.

whatever they could do themselves. When I put the first message up they fought me, saying I needed a sign permit. Then people in health and safety said I don't have to comply with the laws. Public safety notices are ok. I don't know. I thought people would be protecting us, but I found out that's not the case. I didn't know when I worked there. Now we have so many new people in this area now and they don't know. So I just want people to know. People need to know that there is a problem here. People need to know that it ain't gonna be covered up. They try to make everything look nicey-nice, but there is a big problem here. I mean, you don't see any sign anywhere else, do you?

Eric's Story: Up Against the "Power of Capital"

Born and raised on Endicott's North side, Eric, 42, worked at the IBM-Endicott plant from 2004 to 2008 with a subcontractor of IBM called Manpower. His father worked at IBM for many years, and his mom only worked at the plant for a short period in the late 1980s. Eric is now pursuing his master's degree in Media Studies, and his interest in film-making and activism is organically linked and motivated by a more personal struggle. As he put it, "I'm a native son and I lost my mom." His mom died of lymphoma and this had a big influence on his media research project, which is described in more detail below. In brief, he has been working on a film on the IBM spill, focusing on how it personally impacted him and how it has changed the Endicott community as a whole. He refers to his work as work informed by "subjective nostalgia" and he aims to avoid making the film an exposé of the issue. He has titled the film *Juggernaut,* which is a term used by Marx in *Capital Vol. 1* to explain the unstoppable and uncontestable force of capital, or, in Marx's words, "the wheels of the juggernaut of capital" (Marx 1930[1867]:708–9). He explained that *Juggernaut* also symbolizes his own journey, emotions, and his own "character defects, or juggernaut of my own inability to see the truth." He repeated several times throughout the interview that this whole IBM Endicott issue, from the pollution to the collapsing economy, is all about the "power of capital" and representation, not just cancer. "From a personal standpoint, I am certainly much more resonant with the activists in this community, the liberal pro-environment folks, but I am certainly not deep green, you know . . . In a lot of ways much of the eco-discourse falls in the hands of the technocrats because it stays in an arena that is containable, that is not a place for the marginal and the marginalized voices. For me, that was a big thing. Every voice is important."

As an activist working on a degree in media studies and media activism, this resident was the "other" academic in the field who could identify with my struggle to mix my anthropology and advocacy interests.[12] When discussing his stance on advocacy, he often blended his response with a comment on his "method" of advocacy in relation to his media studies thesis project, and how it even differed from what I was doing as a researcher:

> Basically what I find with my method is that it is more about aesthetics. It is different from yours. I have not done one release form. I have done a lot of interviews. Some with residents with a vested interest, some with an activist stance, and some with a grudge. I feel that I am a lot more resonant. I feel that the work that I am doing really contests and blurs the lines between the subjective and the objective. I kind of feel that if people don't like the film, I am just kinda like, "oh well." I feel resonant enough about representing that, of course, and because it is more subjective and representative. I mean I build certain relationships with these people, but at the same time it is very easy to get swayed by people. They are not on

my committee. As much as my heart bleeds for this community and for the people that have suffered, I mean, ultimately, it has got to be my baby. So I find myself having to retrench and rethink the process.

I worked at IBM ful ltime while I was going part time to BU to finish my degree. I was one of the skeleton crew of IBM leftovers. One of my co-workers used to joke saying that "you are one of the rotten people." One of the things that I started to delve into was looking into the culture endemic to IBM and looking at the Endicott plant as being the kind of flagship plant for a long period of time. The ethos that IBM worked under was certainly informed by a Fordist orientation. It had a very strict, austere work ethic. It was very demanding employment. I don't know. I think I resonate a lot with the literature on how collective psyches tend to operate. You know, what kind of things that make up the psyche of an individual can like kind of inform an entire milieu.

Much of his commitment to challenging the subjectivity/objectivity demarcation in his film project spilled over into his discussion of an experience he had at one of the many public meetings that focused on the IBM plume:

I remember going to a public meeting and telling people what I was thinking of doing and I remember that night that everybody was just so staid and so accommodating. It was almost like. . . . I don't know. I wish I had some literary metaphor to make sense of it. It was like being in the land of the lotus eaters.[13] I remember going home that night and calling my cousin, who is with Alliance@IBM, and going, "I have this feeling that I've been had" and he goes, "Well you have" [laugh]. At that point I started recognizing more and more that this issue and the discourses surrounding it were being so denuded of any emotional vitality that it's like . . . I don't know . . . It's like you are walking through a hologram with these discourses, these conceptual models and these agendas that are based on and informed by this technocratic dispassion that had no connection with the lived realities and concrete realities of the people living in this community and suffering from the economic devastation and living with the question mark of living with these inept substances. I wanted to find the heart of it. You know. My feeling was that there was a very distinctive malaise that fell on this community. A malaise that was almost palpable in a lot of ways. I mean that is an aesthetic and artistic judgment about things, but having grown up here, I know how things have been a lie. I am not saying that when the kingdom was in its heyday it was a living lie. At least on the surface it was one of vitality, affluence, and pride.

Eric's activism took root in response to his frustration over IBM's unfair abandonment of Endicott. He likened this theme of IBM abandonment—a central topic of chapter 4—to a "postcolonial" struggle, adding that he felt it was his role to represent the "symphony of voices" that have been silenced and marginalized by IBM power. Eric also wants his film project to be "heartfelt" and "emotive":

> What I hope *Juggernaut* becomes, as much as anything, is a kind of symphony of voices of not just individuals, but of a community and is an exploration of ideas that is unique. I am not doing a cerebral gag. You know. I am too emotionally invested in it. I want to do something that is heartfelt, that is emotive. I am still looking for the heart of the situation. I am still looking. I didn't want to do an exposé, but with all the smoking guns in the situation it became kinda hard not to wanna do an exposé. My contestation, and it is informed by a kinda postcolonial issue, was that this was a culture that had become a world power, you know, IBM, and it had utilized the local infrastructure and the indigenous population to build an empire and when it was no longer profitable to them, they pulled out, leaving the indigenous culture in a bag, you know. This issue is so nebulous. You have to wonder what is left to a community like this.

My goal for interviewing this activist film maker was also to better understand how Eric went about navigating his relationship with fellow activists in the community. I asked him, "How do you talk about your work and the film project with fellow activists in the community? What is it like communicating your interests to them?" To these questions he responded:

> You need to become exceedingly deft in the way you do it. Once again you are dealing with so many different kinds of discursive modalities that you become very deft and chameleon-like in the way that you go about things.[14] At the same time . . . I mean for the activists that have seen the film, I try to explain to them in this sort of laymen parlance without feeling the need

to condescend about what the film is about. Like guys like Larry, he loved it. He came up to me after the viewing and said "Great job. You know it doesn't have to edit the reality." I said to myself, "All right. They get it. They get it." I didn't know how the activists were going to think about it. My view was that if I can do any service to this community I should do a film that doesn't just serve as an advantage on late-night PBS, but we do a film that generates buzz and is kinda controversial. I figured the more buzz there was the more that the issues become known. People might hate it or people might love it, but I'll tell you what, it opens up a discourse, you know. As many Michael Moore bashers there are out there, I'll tell you what, right when people see the film they start talking about the issues.

He went on to add that not only does he feel his film project "opens up a discourse" and gets people "talking about the issues," but it also reveals the humor that Ted tries to keep alive via his message boards: "Part of the integrity of the film is that integrity is not un-synonymous with having a sense of humor about things." But, Eric reminded me that his media activism is something he takes very seriously. Connecting with people is a serious matter for him, especially when, to use his metaphor, you are in "battle":

Once again, I am not trying to get one over on people in this community. It is just like this is the way that I'm doing it, you know. The fact that my mother died I am not, um, I certainly use my mother's case to connect with people. I am not some guy from out of town with some dispassionate need to exploit. I didn't have to go to Chernobyl to do this kind of film. The line of battle starts here. The line of battle against the powers and potencies, the juggernaut that is slowly and imperceptibly running us over, starts right here. This is the front line, you know. It's wherever you are, you know.

This "battle" theme remerged when I asked Eric what he would like to see happen in the community and what his hopes were for the WBESC:

My hope would involve a combination of the veil being lifted a little higher and that you'll never understand the fatalistic aspects of going up against the system. It's a degree of war. It's about the degree of it. It's a battle of shifting boundaries and gaining a little ground that wouldn't have been gained if you weren't there. The mechanism that was in place, you know, and the people that recognized it . . . there was just the right ingredients to start it, with RAGE, CARE, and the WBESC. There was just a lot of organizational talent there that was just fortuitous.

Larry's Story: Retired IBMer and "Lover of the Outdoors"

Larry, 56, grew up in Rochester, New York, and started working at the IBM-Endicott facility in 1973 after he graduated from Binghamton University. As he put it, "I just ended up staying in the area because I got a job with IBM. If you got a job with IBM, you didn't pass it up." I happened to interview Larry on his first day as a retired IBMer. After 35 years as a computer programmer at IBM, and a "computer repair guy" for six years prior to being an IBMer, he shared with me his excitement for his next adventure: "I am moving out west for a while to work as a volunteer in the national park system. I can't wait. I have always been a real lover of the outdoors." Larry decided to join the Sierra Club in 2000 and now chairs the Sierra Club's Susquehanna Group. He also chairs the Natural Gas Taskforce, which targets an issue that is currently the most popular and intensely heated environmental debate of the region: natural gas hydrofracking (or fracturing).[15]

When I asked Larry why he got into environmental activism, he reminded me that sometimes people can get involved in something that they don't actively "seek out":

LARRY: Well, I always loved the outdoors, but didn't really get into any activism until the late 1990s. I realized that well, if I love the outdoors, I figured I should be involved in protecting it. I also realized that if pollution is going on, we need the community to speak up. I am only trying to add my voice to the issue. The Sierra Club stuff has pulled me into stuff that I didn't expect. The natural gas issue has really shifted my work these days. My point is that I didn't really look for this stuff, it just sort of comes along. It's just part of the territory. If you are part of a group like the Sierra Club, you have to see how you fit in . . . You know neither the natural gas issue or the Endicott plume issue is pleasant or something that I would seek out, but you know, if you're gonna wear the badge, you have to participate at the level that you can. . . . I am not an expert. I am just a volunteer.

PETER: So, why did you get involved in advocacy work on the IBM-Endicott plume issue in particular?

LARRY: Well I was involved with the Sierra Club at the time that this
started to really surface and we were not at the forefront of this issue
when it started, but we were kind of watching it from the sidelines
for a while. Then I realized that we had to take a more active part
in it, just to have voices. There were a number of different types of
people involved initially, you know, community members. So I just
felt it needed a more environmental perspective and that the Sierra
Club needed to play more of an active role.

RAGE actually deserves the most credit for really being a catalyst
for getting everything started. They were responsible for getting
the state involved and making a lot of progress. Then the chance to
form this stakeholder coalition [the WBESC] came along and I had
already been working some with RAGE at that point so we all just
decided, yeah let's be part of this stakeholder group. That's really how
I got into it.

I asked Larry to share with me his vision of the future of the WBESC.
He began by sharing with me his understanding of the "limits" of one of
the coalition's greatest accomplishments, which was getting the National
Institute for Occupational Safety and Health (NIOSH) to study former
IBM-Endicott plant workers.

Definitely with the NIOSH study we have already been forewarned
will have limits and that's understood. We do want something though.
Whether there is an increase risk to IBM workers, former IBM workers,
and maybe the study does have the chance to be one of the more comprehensive health studies that come along in the electronics packaging
industry. The fact that there are 28,000 medical records that they have
access to, from a few different sources, gives it a chance for something
really substantial. More than anything else that NIOSH has done before.
NIOSH itself has never done anything like this. So, we realized that
they are not really going to be able to necessarily point to the cause of
increased disease rates, but I do think it's going to speak volumes by just
saying, if they do find increased rates of health issues, something about
the health issues of workers in this industry. So that's a big thing that we're

hopeful for. It's going to take three to seven years, depending on how they go about it. The other big thing in the future is to watch this plume be reduced so it's mostly just a small little part of what it once was. We probably could have influenced them more if we had more people. We have kind of dwindled down in numbers. Initially there were almost thirty people that came to the first stakeholder meeting, and there is just that fall-off rate over time. Now we are lucky to have five people at a meeting.

Larry's final comment here brings up an important issue that seems endemic to social movements. That is, there is a certain ebb and flow of citizen action. Whether we are talking about a local coalition like the WBESC and/or a large social movement like the environmental justice movement, there seems to be a steady cycle of contention in tandem with fluctuating cycles of citizen participation (Tarrow 1998). In other words, there are moments of action and moments of "quiescent politics" (Checker 2005). We saw this issue raised in the other activist narratives presented earlier, but other players involved in the IBM-Endicott contamination debate highlight this trend or pattern. In an interview with a lead NYSDEC official working on the remediation of the IBM-Endicott plume, I was told that "Usually there are only a handful of people in these sites who really get involved and take action. Many of them work their ass off for a long time and end up, for one reason or another, kind of burning out. Sometimes the mantle is picked up by somebody, but in many cases I think just the people that are the energy behind it burn out and involvement starts to dissipate. I think Endicott is kind of following a normal pattern that way." In the case of Endicott, the mantle has been picked up by Shelley and other members of the WBESC, especially Frank. On the other hand, many activists talk about this theme of dissipative activism, expressing their concern that they are seeing a decline in public involvement.

After a coalition meeting on a cold and snowy winter evening in January, one WBESC member told me as we walked to our cars: "You see how many people made it to the meeting? These rooms used to be packed with people. Now no one comes. They don't seem to care." He added that one reason for this is because a lot of the plume residents are elderly and it is too difficult for them to make the meetings, which are usually held after dinner time, or around seven or seven thirty. One plume resident I interviewed, who had attended many of the first public meetings, explained to me why she stopped going to meetings: "It is like every time the same people do all the talking. The people who run the meeting do all the talking and take up all the time for questions. It is really a circus most of the time, so I stopped going. It is just a waste of time most of the time."

When asked to talk about why he thought meetings were seeing such low turnouts, he told me: "We have just a whole bunch of new people living here who just don't care because they are living here for free. It is all Section 8. We have people moving up from New York City. That's what is going on here." While not using these exact words, most every-one I interviewed mentioned that ever since the 1980s, the area now known as "the plume" has shifted from a population of homeowners to a renter population. For this WBESC member and plume resident, people are not coming to the meetings because the majority are rent-ers. While my research did not investigate the degree to which home-owners or renters compare or differ when it comes to how much they "care" about the IBM contamination debate, my survey of plume resi-dents did show that more homeowners than renters attend public meet-ings on the issue. For example, my survey results indicated that 74 per-cent, or 37, of the homeowners surveyed (n=50) had attended public meeting(s), while only 16 percent, or five, of the renters surveyed (n=31) had attended a public meeting(s).

According to Mike—a WBESC member who was discussed earlier—declining attendance at meetings and the slow dissipation of public involvement could also be a sign that things are actually going well, that the WBESC is "doing a good job":

> I'd like to think it shows that we have been successful. Our meetings are further apart now because the bulk of the work has been done. Now we only have meetings when it is necessary. It's not monthly meetings any-more. Now we have to look at what we are going to do in the future. The two things are the workers' study and the remediation issue. Those are the two major things. As far as people coming to meetings now, I don't see a lot of people coming anymore. If people were really interested they would come. People know when the meetings are. I don't want to say that I don't care that they come or not, but if they were really interested they would be coming. The other way to look at it is that they think we

are doing a good job and so they don't need to participate. I don't know that that's true either.

Witnessing a Kind of Boundary Movement

When shown together as a collection of activists' narratives, what do these repertoires of action and intentionality represent? Grassroots action in Endicott links to theoretical debates in environmental social movement studies, particularly those engaging theories of movement hybridity, or what some researchers have called "boundary movements" (McCormick et al. 2003). Boundary movements have been described as "social movements . . . [whereby] their constituent organizations . . . move between social worlds and realms of knowledge. In so doing, they blur traditional distinctions, such as those between movement and nonmovement actors, and between laypeople and professionals" (ibid.:547). Boundary movements, according to these researchers, have several key characteristics. "First, they attempt to reconstruct the lines that demarcate science from nonscience. They push science in new directions and participate in scientific processes as a means of bringing previously unaddressed issues and concerns to the attention of the clinical and bench scientists" (2003:547). A second quality of these movements is that they "blur the boundary between experts and laypeople" (ibid.). By this they mean that "some activists informally become experts by using the Internet and other resources to arm themselves with medical and scientific knowledge," while "Others gain a more legitimate form of expertise by working [directly] with scientists and medical experts" (ibid.).

A third feature of boundary movements is that they "transcend the traditional conceptions (i.e., boundaries) of what is or is not a social movement. They do this by moving fluidly between lay and expert identities, and across various organizational forms" (ibid.). Finally, these movements cross "two or more social movements, while blurring the boundaries of those separate movements" (ibid.). There are a number of reasons why I consider citizen action in Endicott an example of a "boundary movement," but these last two characteristics seem to fit best because the line between "occupational" health concerns and "public health" concerns is blurred among these activists. I found that vocal activists and non-activist residents alike generally consider the

IBM-Endicott pollution debate a problem of both workplace and home exposures. Exposure risk talk, according to those I interviewed, did not agree with the boundary made between work and home, despite mandated government agency efforts to demarcate the two spaces of exposure and "treat" or manage the risks posed in these spaces differently (e.g., with different scientists, different regulators, different regulations and standards, etc.). Most engaged local activists are aware of these agency-based jurisdictions—ATSDR is responsible for "public" health investigations and NIOSH is responsible for "occupational" health investigations—but many of the residents I interviewed don't understand why home and workplace exposures are not joined and studied in synergy to make sense of the broader "community-wide" contamination problem. While uncertainty, invisibility, imperceptibility, and doubt are themes that commonly surface for residents living in contaminated communities (Edelstein 2004), and certainly are themes playing out in Endicott, the WBESC has remained certain and driven to keep the work they do open and, in a sense, "boundless" to some extent. I say to some extent because one coalition member expressed to me the need to keep a particular focus to stay organized, which meant avoiding other environmental issues in areas such as natural gas drilling and hydrofracking of Marcellus Shale (Wilber 2012), which has become a matter of concern and action for at least one of the WBESC members I interviewed.

Checker (2005) offers a point of view that is similar to that argued by boundary movement theory, contending that environmental justice movements, like all social movements, are "messy," in that their "dynamics are overlapping and unruly. They defy boundaries, models, and temporal isolates" (Checker 2005:34). It is often difficult to pin down what is "in" and what is "out" when discussing struggles for justice that stress problems or "rights" that are very broad in nature (e.g., human rights, rights to a healthy environment, and rights to health). This flexibility has opened up the vision and possibilities of certain social movements, like the contemporary environmental justice movement, which is currently experiencing an intensification of its vision of justice and who in fact qualifies for these struggles for justice:

> The solution to environmental injustice lies in the realm of equal protection of all individuals, groups, and communities. No community, rich

or poor, urban or suburban, black or white, should be made into a sacrifice zone or dumping ground . . . The environmental justice movement
challenges toxic colonialism, environmental racism, the international
toxics trade, economic blackmail, corporate welfare, and human rights
violations at home and abroad. Groups are demanding a clean, safe, just,
healthy, and sustainable environment for all. They see this not only as
the right thing to do but also as the moral and just path to ensuring *our*
survival (Bullard 2005:42; emphasis mine).

But while this movement transformation is under way, certain activists
struggle with the very idea of couching their struggle in terms of environmental justice. My ethnographic research found that some toxics activists
wonder "where environmental justice fits" or "where the environmental
justice connection is." These are provoking questions that confound the
general understanding of environmental justice as "a local, grassroots, or
'bottom-up' community reaction to external threats to the health of the
community, which have been shown to disproportionately affect people
of color and low-income neighborhoods" (Agyeman 2005:1–2).

As mentioned earlier, local citizen response has *enabled* both a "sense
of community" and a "politics of enunciation" (Allen 2003, 21)[16] show-
casing at least some sense of collective intentionality and environmental
morality. What I came to realize when I returned to Endicott in 2008, and
what I hope to expose in the ethnographic detail that follows, is that augmenting this "politics of enunciation" is the discomfort and disconnect
between local activists and outsider environmental justice activists, both
of which, and I include myself here, tend to "make it their duty to speak
for the people" (Bourdieu 1990, 186).[17] But, what if we also made it our
duty to "speak for" the micropolitics of alliance-building, the vibrant and
yet messy inter-subjective relations between grassroots activists?

Conflicted Environmental Justice

The number of studies examining grassroots conceptions of environmental justice has been steadily growing over the years (Allen 2003; Bullard
1994; Cable and Shriver 1995; Capek 1993; Checker 2005, 2012; Di Chiro
1995, 1998; Harvey 1996; Kurtz 2002; Lake 1996; McGurty 2000; Novotny
1995; Pulido 1996; Sandweiss 1998; Schlosberg 1999, 2004, 2007; Spears

2006; Szasz 1994; Taylor 2000) and it is safe to say that the contemporary social identity of the environmental justice movement is more nomadic and in-the-making, than it is fixed or firmly decided. So too are theories of environmental justice. As Schlosberg (2004:517–18) puts it, "most theories of environmental justice are, to date, inadequate. They are incomplete theoretically, as they remain tied solely to the distributive understanding of justice— undertheorizing the integrally related realms of recognition and political participation. And they are insufficient in practice, as they are not tied to the more thorough and integrated demands and expressions of the important movements for environmental justice globally." Furthermore, while the contemporary environmental justice movement has been mediated by other "repertories of action" (Hess 2007b), other "projects of representation" (Marcus 1999), this in turn has complicated ethnographic studies or accounts of environmental justice struggle because these ethnographies end up amounting to "the disentangling [of] dense webs of already existing representations" (1999:23).

What follows is an exploration of local activists' contentious uptake of environmental justice. I consider this a confrontation with *conflicted environmental justice*. What I am getting at with this phrasing is something that is, according to Sen (2009), "central to the idea of justice": "we can have a strong sense of injustice on many different grounds, and yet not agree on one particular ground as being *the* dominant reason for the diagnosis of injustice" (2). Sen calls this the "plural grounding" of justice. While the need to prove or disprove environmental justice is for many environmental justice activists a low priority and even counter-productive to the counter-hegemonic aims of the environmental justice movement, the politics of self-identification start to matter when the movement becomes a movement "for all." Here, my focus is not only the "plural grounding" of environmental justice, but also the contested grounding of the local diagnosis of and confusion with the very *existence* of environmental justice politics in Endicott. In other words, I am drawn to the micro-level ruptures that emerge with and even in response to the expansion of solidarity movements like the environmental justice movement.

When perplexed about environmental justice and the legitimacy it has as a framework and cause for action in Endicott, when asked to think differently about their toxic situation, what do people say? This question has guided what I have called elsewhere the "micropolitical ecology of environmental justice" (Little 2012b). While activists are often sure about why they act, why they engage in community advocacy and become vocal stakeholders, there are events, perspectives, and discourses that travel and move into contaminated communities from "outside" that can invoke critical contemplation. One such event and traveling discourse was the National Environmental Justice for All (NEJA) tour that came to Endicott in the fall of 2006.

The National Environmental Justice for All Tour

In May 2004, environmental justice organizers from Los Angeles, Louisville, New York, New Orleans, and Texas gathered in Louisville, Kentucky—together with other advocacy groups from across the country—to support the grassroots work of Louisville activists who had been struggling for many years to force a cleanup of industrial pollution in a section of West Louisville known as "Rubbertown." Twelve plants along a half-mile stretch polluted an adjacent African American neighborhood on a daily basis, resulting in activists from around the United States descending on Louisville to show their solidarity and help explore available solutions. This collaboration, led by local organizers, eventually paid off. In June 2004, the Louisville Metro Council passed a budget to fully fund a Toxic Air Pollution Reduction Plan. In June 2005, the Council passed the Strategic Air Toxic Reduction (STAR) program, a regulatory program designed "to reduce harmful contaminants in the air we breathe, to better protect citizens' health and enhance quality of life."[18] Although great successes were achieved in Louisville, the STAR program is under attack by big polluters and activists still work tirelessly to ensure pollution reduction goals are made a reality.

Organizers soon realized that Louisville's struggle was common in communities across the country and that EJ activists were networking and joining together to augment the EJM. Faced with this realization, EJ organizers at the 2004 Louisville meeting brainstormed a National Environmental Justice For All Tour to highlight EJ struggles and "meaningfully link community groups together." According to these organizers, the goal is to raise "the profile of EJ fenceline struggles at the same time as building solidarity and influencing major chemical policy reform debates with the experiences of grassroots fenceline communities."

The tour organizers connected their cause and method of action to former and more recent Freedom Rides. The 1961 Freedom Ride began in Washington, DC, and was destined for New Orleans. Although the African American and white Freedom Riders never reached their goal because of beatings by mobs of white racists in Alabama and Mississippi and arrests on trumped-up charges, they achieved important civil rights victories in the form of federal laws that outlawed racial segregation in interstate travel and accommodations. More recently, in 2003, the Immigrant Workers Freedom Ride successfully employed this civil rights strategy to advocate for a legal status for immigrants who continue to face exploitation despite living, working, and paying taxes in the United States. Building on the legacy of the Freedom Ride, the Environmental Justice for All Tour hoped to create momentum for reforms of governmental, corporate, and institutional policies and practices that result in environmental racism, expose people to toxic pollution, and damage our environment.

The goals of the tour illustrate the "new pluralism" of contemporary environmental justice movements (Schlosberg 1999). While organizers aimed to ensure that the tour would inform policy initiatives at the state and national levels, that it would augment the political base for environmental justice work and impact leadership in the 2006 congressional elections, the tour also aimed to "Bring together multiracial and multiregional grassroots community activists, environmental, health, and social justice advocates, medical professionals, policymakers, celebrities, and journalists aboard biodiesel buses, fostering solidarity among these diverse groups."[19] Environmental justice *for all* was the goal. What is important, I think, is how such universal environmental justice ideas and practices are themselves appropriated and re-embedded in and by local activists and their practices, thus exacerbating the fragmentation and messiness of environmental justice action and the environmental justice movements' attempt to proliferate and collectively organize. What follows is an ethnographic sketch of local activists' interpretations of the NEJA tour and their negations with environmental justice.

Micropolitical Narratives of Environmental Justice

A central focus of the ethnography of environmental justice is to develop an understanding of the significance that certain concepts, such as advocacy and environmental justice, have for people engaged in citizen action. In this way, the narratives I collected expose how residents and activists reflect on the EJ Tour and negotiate how and the extent to which the EJ framework fits or figures in local narratives of toxic struggle in Endicott. As I have argued elsewhere (Little 2012b), these activists' narratives also illustrate certain micropolitics of local advocacy, setting in relief the converging and diverging views of WBESC activists and their sense of discomfort and disconnect when it comes to environmental justice activism. Some of this discomfort and disconnect, as the narratives reveal, is informed by deeper racial politics that at least in part stonewalled and disable effective alliance between WBESC members and members of the NEJA tour.

To be fair, it is not as though activists in Endicott outright contest the environmental justice connection in Endicott, but in their narratives on making the connection, and reflecting on the NEJA tour, the discourse seems, to me, struggle-like. Perhaps it is easier to locate what activists are committed to, what they are sure of, than what is confusing and ambiguous. The topic of environmental justice can invoke this sense of confusion and ambiguity, even for more Leftist activists in the community who are privy to *justice* struggles. This came to my attention especially in my interview with Frank, a union organizer with Alliance@ IBM and active member of the WBESC. I draw special attention to his narrative here, for he spoke at great length about the NEJA tour.

Frank sees the NEJA connection mostly in terms of a problem of widespread corporate pollution. According to him, there are many other communities like Endicott dealing with "this kind of thing," but maybe not necessarily the contentious and racialized zoning decisions for toxic industries, which usually invoke environmental justice-oriented politics and forms of resistance. He welcomed the tour and its cause. He felt it helped show that the issues in Endicott go beyond Endicott, that corporate pollution affects "all people," even though he also felt the NEJA tour was, as he put it, "very strange." For Frank, the IBM spill was an environmental issue at first, but has become an environmental justice issue, though it took the NEJA tour for him to realize this. He explained to me that he "felt weird" when the NEJA tour came to town because the majority of Endicott residents are white. He admitted to me that he was confused at first and didn't get why they were in town. At the same time, Frank felt the tour helped him rethink how the IBM pollution in Endicott compared to other places:

> In terms of the environmental justice tour, I think that this area definitely applied because there are many other communities dealing with this kinda thing. The tour was quite a large undertaking. It was done on two coasts. One on the west coast and one on the east coast and I think they had planned to do something in the central U.S. They had a tour on the east which Endicott was a part of. There were many communities on the tour where racial inequalities and the zoning of factories and pollution was a factor. We didn't really recognize that here in Endicott for one kinda weird reason and that is that the majority of Endicott residents are white. I don't know the reason for that either, but it just happened to be that the majority of the people affected by the pollution in the plume and the workers were white. They were just not African American or Hispanic numbers in the area. I am sure there are people here who are either African American, Hispanic, or Asian or whatever, but the numbers were not very large and they were not the people really affected by the plume. But, when the tour came they brought with them several people that came from communities that were primarily African American or Hispanic and basically non-white. So their issue had that additional element.

Frank felt the tour gave the activists in the community "a bigger view" of the problem: "it gave us a bigger view. It showed us how this doesn't just happen to people and communities like Endicott." On the other hand, he also explained that all the members of the WBESC and RAGE are white and the primary reason they took action was because they felt that IBM "was picking on them or that IBM was disregarding them as people." He added that "It was a very interesting perspective that the tour activists took. I hadn't even thought about it being a racial perspective until the environmental justice tour. I just said 'Wow. This is very strange.'" What added to the experience of strangeness, at least for Frank, was that at the public meeting held to welcome the NEJA tour, "one of the African American gentlemen got up and spoke and he was very preach-like, with kind of the Southern Baptist kind of delivery." Frank cautioned that he didn't know if he was Baptist or not, but that "it was just that kind of delivery." He continued, though with some trepidation, that "I happen to know personally that there are some racial problems with some of the people I have dealt with between the WBESC and RAGE, and some of those people were very uncomfortable hearing this guy talking about injustice in other communities."

Frank wanted the tour to expand the vision and mind-set of people in Endicott. He felt that residents needed to understand that the issues in Endicott are happening elsewhere and affecting "all people." For him, that is precisely what the NEJA tour was "all about." He also shared with me his "impression" that some comments and actions of certain RAGE and WBESC activists and residents he knows and works with have been fueled by an undercurrent of local racism and prejudice toward minorities, which he finds ironic because Endicott has historically been a community of immigrants: "It affects all people. It doesn't just affect former IBMers and the Village of Endicott, or just people who used to work for the Endicott-Johnson Shoe Company [locally known, oddly enough, as EJ], or people who lived here for eighty years plus who happened to be all white." Many of these EJ workers he refers to were Italians, mostly Sicilians, who immigrated to the United States in the late 1880s and early 1900s to work at the shoe factory. Today, Italians remain an ethnic majority in Endicott.

When the NEJA tour came to town, Frank "kind of got the vibe" that some of the people and activists he was involved with were "uncomfortable." "I am not going to call them racist because I don't know what they believe in their hearts and souls. But, some of their actions and some of the things they have said and done since the tour kind of gives me the impression that they are." After telling me that he gets accused all the time for being a radical Leftist and communist union organizer, he added:

> The tour was about participation and about making the participation nationwide, not just in one community. I mean we really tried to link up to other communities after this tour to get a dialogue going between the communities and organizations and basically disseminate that to the different communities. Both the WBESC and RAGE felt that this tour was a good thing and that we could help each other. We could get help and help each other. It was just a larger perspective. We expected something to come of it. Whether or not anything has come from it is difficult to say. I am a cynical person. I am glad it took place, but I am not so sure that it was what it should have been, with the results and the outcome of the tour.

Frank was disappointed with his fellow residents and activists, and he is agitated by the "enormous amount of fear," of people being afraid to "open their mouths," especially former IBMers who fear they will lose their pensions. Ignorance, he explained, is another point of frustration:

There is just an enormous amount of fear. Here and in other communities. Nevertheless, it seems to be very prevalent among the community. I can't tell you how many meetings I have gone to since that tour that you just leave with the impression that people are afraid to step forward and many of the people who don't think it is a problem were either on IBM's side because they are a part of it or because they are simply on IBM's side because they don't know any better or they are not necessarily on IBM's side, but they don't believe the pollution is a problem because they are ignorant. Sorry, but that is basically it. Those are the people who come out and speak against it and it scares the living day lights out of some of the people who want to open their mouths.

Eric, who was born in Endicott and worked at IBM for a short time and is now earning his graduate degree in media studies, found the NEJA tour completely appropriate and spoke to the real "forces at work," "the forces of globalization" in deindustrialized communities like Endicott that are "witnessing the backlash of the northeast metropolis." From his perspective, the tour was "fortuitous" in the way it set in relief this broader political economic narrative:

I think that the tour was fortuitous in the sense that it informed a local grassroots movement with more overarching and national issues that might have not been there otherwise. Again, we talked about how everything is political and everything is social and economic, and the kind of forces that I see at work in this particular community when it comes to the allocation of people in lived environments who will be the residents of this kind of toxic hot zone and how one keeps the tax base from falling out, I see it as almost as a science fictional landscape in which the lower class, people of color, the mentally infirm, the elderly, the lowest echelons, not being armed with the information they need will be sequestered into toxic environments where nobody else will live in. I see that as becoming more and more of a reality here. Of course, Native American lands were targeted. But, what you see in downtown Endicott now and due to the forces of globalization is witnessing the backlash of the northeast metropolis. We are certainly seeing the sequestering of people of very low needs and here, I mean, most of the people that lived in those homes in the plume were working class and experienced a dive bomb and moved out, you know, and the lower class is moving in.

The demographic shift alluded to here is a local factoid that speaks directly to environmental justice. All the residents and activists I interviewed mentioned that the population of renters in the plume area has been on the rise since at least the late 1990s. Members of RAGE and the WBESC helped prod the state to address this issue and make sure that the renters were being properly informed about the TCE plume they lived above and the "mitigated" home they were renting. These activists got what they pushed for. In 2008, New York State assemblywoman, Donna Lupardo, helped pass a tenant notification bill which aimed to "require property owners to provide notification of testing for contamination of indoor air to current and prospective tenants." It is listed as an Environmental Conservation Law, but it is unclear which state agency is responsible for enforcing it. Many residents and activists like Eric believe it is just something on paper: "This tenant notification bill isn't doing anything. I question the power of the people and their ability to even have the conceptual apparatus to process the law anyway." In fact, results of my household survey of 82 residents (both renters and homeowners) living in the plume, which was completed several months after the tenant notification bill was passed, indicate that no renter surveyed (n=33) knew nothing about the law. Tenant notification and rights to information are surely known topics of environmental justice in contaminated communities, but as far as I know the former has yet to become an explicit stimulator of local grassroots action in Endicott.

Not all local activists interpreted the NEJA tour as Frank and Eric did. For example, when I asked Diane, a former member of RAGE and the WBESC who is now a New York State assemblywoman, to talk about the NEJA tour, she was more forward about Endicott *not* being a legitimate community fitting within the environmental justice frame. Environmental justice is just not on her "radar":

I have certainly not framed it this way locally. I don't think many local residents are thinking in these [environmental justice] terms. Certainly as a state legislator I have worked with colleagues from the Bronx and other places that are really communities of color and those impoverished areas where facilities have been specifically sited and where groups have been truly taken advantage of through their lack of education. For me, I am a sponsor of the environmental justice bill and one of the criteria that seems to be holding up with the zoning of power plants is the environmental justice issue and providing communities with the knowledge, awareness, and tools to assess the impact of these things on your community. Um, this was a company town which either through neglect or through mistakes have led to the contamination of the groundwater so I guess I don't see it in those terms, no. But, I also may have a more narrowly defined view of environmental justice in this country. The way I think about it, environmental justice is not on my radar here. It's not what's going on here.

Some WBESC members did not see the tour in this light, and feel they *do* have a confident "conceptual apparatus" and are sure Endicott and the goals of the NEJA tour don't mix. For example, Shelley, who self-identifies as an "outspoken Republican" and is the present leader of the WBESC, informed me that she was "kind of reluctant" when the NEJA tour came to town:

When I heard the environmental justice tour was coming through I was kinda reluctant. There is a lot of propaganda in this issue, a lot of propaganda. All I knew was, they asked me to speak and I did, but I also prepared a little paper, as best I could, as a handout. I felt a little better getting up there and speaking to the group, but you know I think I saw it for what it really was, and it was a propaganda type of thing, I felt. It was geared toward a more Democratic agenda and that is what it really was. I don't think it has too much to do with environmental justice as it has to do with promoting one party's agenda. I agreed to be a part of it because if I wasn't there then the WBESC wouldn't have been there and there would not have been any reality to it.

She then added that de-politicizing the issues was important to her and other WBESC members, and the NEJA tour threatened this goal:

> I think I even mentioned to the NEJA activists and the WBESC members that this was not a political issue and that we had to all do this together. I felt kinda strange and this may sound racist, but I don't care. They had a couple of black people who got up and spoke and said that we have black people in this community who don't have any rights and don't have any opportunity. That is false. Everybody in this community has an opportunity if they want to apply themselves. That is a basic philosophy I have and when you hear they are coming up from the South, they are still promoting those kinds of agendas and I don't think that is healthy for this country. I think we have to still empower people, not by entitlement, but self-responsibility.

This is the discomfort that Frank spoke of earlier. Shelley added that she didn't really see the environmental justice connection because when she visited "the South" for a family vacation, she didn't "see" the racial problems and conditions of injustice underpinning environmental justice struggles and the issues and concerns that NEJA participants spoke of when they visited Endicott. "I didn't see it down there. I asked my husband, 'Where are all these problems? Where are all the racial problems?' I didn't see it. What I saw with the environmental justice tour was a turnoff, you know. I didn't see it." "So, you know, whatever color you are, whatever your political persuasion, if you don't want to help us, if you don't want to work . . . I'm done. I don't have to apologize for anything," she added.

Some WBESC members I interviewed find Shelley's actions and attitude a little harsh. Despite his belief that Shelley's candid personality can make him uncomfortable, Mike, a WBESC member and retired engineer, interprets the NEJA tour in much the same way that Shelley did, though with much more modesty. Before sharing his perspective on the efficacy of the NEJA tour, he reminded me that it was unlike him to even get involved with an environmental group like the WBESC. "I don't normally get involved in these kinds of groups. I think one of the reasons that I did get involved was more or less associated with the state health department . . . I guess I didn't want to get involved with an environmental type of group. The group was more, I don't know, government sponsored. That impressed me." He felt that the government agencies were going to listen to the coalition and so he decided to join. As he put it, "I didn't become a member for really any other reason. I have never become a member of a group that would advocate in a radical manner. So, when the Environmental Justice group came, I would say that they have some good points, but I would not totally agree or think that that is the way to go about it." Mike explained that the WBESC has direct access to the agencies and that they can "get someone to be here to address the issues." He felt this was something that differed from other organizations that seem to struggle for attention. Reflecting on the NEJA tour, Mike stated that "with these other groups, I guess I don't see the value as much, as much as the stakeholders' [WBESC] group."

Larry, a former IBMer and the WBESC's Sierra Club connection, pointed out that Endicott was "a little unique for the tour," a perspective that resonates with Frank's position. When I asked Larry to talk about his experience with the NEJA tour, he responded:

There is not a real large minority population, but there is a fairly large poor demographic group that lives in the plume area. By and large, both people haven't been involved. Maybe that is a limit or maybe we are lacking in that, because it seems to be the case in other parts of the country. Endicott lacks minorities. We are lacking in that. But, I mean environmental justice doesn't only have to be dealing with minorities. Women are somewhat disenfranchised. Some people don't have political clout or power and that is definitely the class of people that were in Endicott at the time. I think the state probably didn't really pay close attention to us for a long time. I mean, if it had happened in a nicer place, they might have done more to clean it up early on. But, they sat on the whole problem for 25 years, maybe with the effect that people would not give too much voice to it. The community found their voice. The first meetings were packed meetings. The state was clearly pushed and they knew they had to respond. But, it's a poor community.

I think Larry is right. Endicott is a poor community. It is a deindustrialized and defamed community. It is like many other Rust Belt communities coping with the aftermath of capital flight, community disinvestment, and other processes of neoliberal creative destruction (Harvey 2007). What seems less clear to the activists I just mentioned is the place of environmental justice politics in Endicott; citizen action rooted in environmental justice motives. CARE, RAGE, and the WBESC have all helped the community "find their voice," as Larry put it.

Environmental Justice Unbound

Where is the voice of environmental justice? Who is "speaking for" or against environmental justice in Endicott and beyond? What is the tone of the micropolitics of environmental justice in IBM's tainted birthplace? These are tough questions that I think inspire environmental anthropologists and political ecologists to attend to critical dynamics of intra-community and intra-activist relations. While the narratives on activists' interpretations of the NEJA tour help to answer these questions, the point is that environmental justice *is* in fact locally negotiated and contested, even amid an expanding social movement informed by a righteous intention to make environmental justice *for all*.

Many activists I interviewed talked about the NEJA tour as an effort and event that created a wide-angle view, a broader perspective on community toxics issues that transcends racial and ethnic boundaries. The narratives described here attempted to show how these activists craft their own grassroots perspective on environmental justice and the meaning and efficacy of the NEJA tour. These narratives exposed personalized visions of the distinction between *us* and *them*, between Endicott's toxics issues and those in focus for the NEJA organizers and participants. Some local activists see legitimate connections. But, there still was an enduring suspicion about what was actually shared. As Frank's narrative highlighted, many WBESC and RAGE activists felt a certain discomfort with NEJA activists and that while he could not say for sure, this discomfort seemed to him rooted in racial politics. Prodding this topic as a former resident, activist, and researcher was a serious challenge. While the narratives analyzed in this chapter aimed to expose these race-relations issues and the social terrain of negotiation, the decision to discuss these tensions, especially in a community that has made great progress in demanding IBM and the state to clean up contamination, was not an easy one. My intent has been to *not* edit the reality of the micropolitics of local advocacy and instead enunciate what was witnessed; to honor what was said about NEJA activists and their efforts to communicate the meaning of environmental justice *for all*. Invoking discussion of the micropolitical ecology of environmental justice in IBM's tainted birthplace, this chapter has tried to expose a situation of environmental justice alliance interruption whereby a movements' efforts to attain a politics "for all" is met with, unexpectedly, local contestation, discomfort, and dis-connect, and even alliance-crippling racial politics. After all, advocates are "situated in reciprocal relation to other advocates, even if geographi-cally distant, whose intended as well as unintended actions influence what is perceived as good and possible" (Fortun 2001, 16–17).

The NEJA tour, while not its explicit mission, attempted to expand the boundaries of environmental justice and reinforce two overlapping poli-tics that continue to inform contemporary environmental justice activism: the antitoxics movement and the movement against environmental racism (Schlosberg 2007, 46). Several studies have exposed this boundary break-ing and overlapping tendency of environmental justice action (Adamson, Evans, and Stein 2002; Ageyman, Bullard, and Evans 2003; Agyeman and Evans 2004; Brown 2007; Brown et al. 2004; Bryant 1995; Bullard 1993, 2005; Cole and Foster 2001; Faber 1998; Hofrichter 1993; Pellow and Brulle 2005; Roberts and Weiss 2001; Stein 2004). Despite some recent efforts (Allen 2003; Checker 2012, 2005; Ottinger 2013; Pirkey 2012;), there is a need for greater ethnographic description illustrating the complexities of intra-activist relations in general and peoples' engagements in and experi-ences with the micropolitics of environmental justice alliance-building in particular. Of course, it is unclear to what extent ethnography can inform citizen action and reconfigure the ecopolitics of "technocapitalism" (Suarez-Villa 2009) unfolding in IBM's birthplace community. For now at least, the meaning and racialization of environmental justice in this socio-political setting is under negotiation, as are the always emerging actor networks linking Endicott activists to other activists, other ecopolitical imaginaries. As a member of this networked terrain of advocacy—even if most often from a distance—I think certain environmental injustices, despite some activists' claims to the contrary, have occurred and continue to unfold in Endicott. The community is already remembered for helping launch the Computer Age and a culture of techno-fanaticism, innovation, and "planned obsolescence" (Slade 2006). It is only a matter of time before the community is remembered as a site of vibrant ecopolitics, a setting of conflicted environmental justice and precarious alliance politics.

7

Citizens, Experts, and Emerging Vapor Intrusion Science and Policy

When the ship is sinking, the fights between the helms-
men above the gangway and the engineers below among the
engines become ridiculous, stupid, and dangerous. In bad
weather, it is better—no, it is necessary—to unite to secure
the ship above and below the waterline.
—Serres (2011:83–83)

Each community has experiences that are worth sharing,
even if they are not quantifiable or easily comparable—the
key is to get their stories out so that others may learn from
them.
—National Academy of Public Administration (2009)

As an anthropologist and political ecologist engaged in vapor intrusion
(VI) debates, I am interested in how the lay public and scientists and
regulators come to know and understand these emerging debates. VI is
one of many emerging sciences and technologies in which anthropology can intervene (Downey and Dumit 1997). It is a risk debate largely
because of the "co-production" (Jasanoff 2004) of scientific knowledge
and state intervention. According to Lenny Siegel, a leading VI activist and technical advisor to communities like Endicott coping with
VI, there is a "rocket science" problem, as he puts it, that leaves many
impacted residents confused and unsure about VI risk, exposure, and
mitigation. How scientists talk about the challenges and complexities
of VI is something that I, as an anthropologist and activist who has
been identified as a "community stakeholder" in Endicott, am drawn to
because I want to better understand the practice and epistemic anchor
points of VI science to better inform myself and other activists and

community stakeholders engaged in VI struggles. In this way, my inter-est in doing engaged anthropology, in experimenting with "engaged political ecology" (Batterbury and Horowitz, forthcoming), in showcas-ing "*scholarship with commitment*" (Bourdieu 2003:24; emphasis his), is born from my eagerness to both learn about something complex like VI science and policy, and empathize with residents who are often genu-inely confused (like me) about the scientific knowledge and risk calcu-lations shaping VI decision-making. I use this chapter to open up VI debates to the sensibilities of transdisciplinarity, of anthropology and political ecology, and to new interventions and possibilities for emerg-ing VI policy.

Accounting for the Complexities of Vapor Intrusion Science

Intrusion rings with the tone of contention, conflict, aberration. *Vapor intrusion* or *toxic VOC intrusion* intensifies the semantic weight by an order of magnitude. In other words, if intrusion emits something nega-tive, "toxic" VI is synonymous with something really bad, really vola-tile. In fact, when I first heard that VI was the focus of concern at the IBM-Endicott site, I remember thinking that the term *intrusion* seemed synonymous with invasion, a word with a corporate colonial blood-line. It continues to symbolize for me how VI is yet another form of toxic appropriation (Serres 2011), a process of toxics exposure that has remained *after* IBM deindustrialization. Like mitigation systems mark-ing the Endicott landscape, the "risk" of VI is front-and-center, as are the always swarming politics of risk. With the help of techno-science and the state, VI risk was *made* public and it was and remains an elusive public health threat. It is an "exposure pathway," as VI experts explain, that is difficult for both citizens and scientists to fully understand, even with sophisticated sampling techniques and modeling tools.

In one interview with a vapor intrusion expert, it was brought to my attention that many challenges come with studying the VI pathway, challenges that can be "rewarding" and exciting from a scientific prac-tice perspective:

> From the technical perspective VI is a very challenging path to under-stand. The groundwater path has its challenges, but it is relatively simple

in terms of contaminant migration through aquifers. You can put in a relatively small number of wells and characterize the nature and extent of contamination that way and feel fairly confident that you can make good decisions. What we have seen with vapor intrusion is that the distribution of VOCs in the subsurface and beneath peoples' houses and then even the VOCs getting into their houses is much more variable. There is a lot of spatial variability in the distribution and temporal variability in terms of changes through time. They are much greater than a classic groundwater plume. So that presents some of the technical challenges of VI in terms of control and understanding the nature of the vapor contamination. It has been a challenge as well as a real rewarding experience. You feel like you are kind of at the cutting edge of what's going on.

In other words, there are many unknowns when it comes to the science of investigating, evaluating, and mitigating exposures related to VI. While measuring indoor air concentrations generally produces the most accurate information for determining *exposure point concentrations* (or what people are actually inhaling), VI scientists and engineers generally use what they call *attenuation factors* to formulate an estimation of the amount of vapor that might be entering the building.[1] While attenuation factors are usually determined by investigating and comparing the relationship between groundwater contamination and indoor air concentrations, it is generally more reliable to use soil-gas-to-indoor-air attenuation factors. Attenuation is a central buzzword in VI discourse, marking a very real challenge for scientific *determinations* of the presence and severity of vapor intrusion risk for residents living above groundwater sources contaminated by volatile organic compounds (VOCs). According to one EPA report, "VOC concentrations in soil gas attenuate, or decrease, as the VOCs move from the source through the soil and into indoor air. The extent of attenuation is related to site conditions, building properties, and chemical properties, and is typically quantified in terms of an attenuation factor defined as the ratio of indoor air concentration to source vapor concentration" (USEPA 2012:1). The migration of TCE vapors, in other words, from the contaminated groundwater source at the IBM-Endicott site is a complicated migration shaped by multiple factors: "[S]everal factors (e.g., subsurface and building conditions) work together to determine (1) the distribution of VOC contaminants in the subsurface and (2) the indoor air concentration relative to a source concentration. Factors [can] . . . include vapor source characteristics (e.g., concentration, size, location, depth), subsurface conditions (e.g., soil layers, moisture conditions, oxygen levels for biodegradation), and building characteristics (e.g., foundation type and condition, pressurization, air exchange rates), as well as general site conditions (e.g., wind, ground cover)" (2012:1).

More accurate and reliable sampling technologies for VI are also emerging. Siegel has recently pointed out that "sampling technologies, using sensors that can be pointed at potential pathways and sources, are a superior way of resolving where VOCs are coming from. To some degree those technologies exist now, but the entire field of vapor intrusion investigation will be greatly enhanced when chip-based sensors become available in the not-too-distant future" (Siegel 2010). Even if these high-tech VI sampling technologies become available at every VI site, exactly how these technologies will benefit affected residents coping with health concerns and property devaluation is unknown.

The complexity of vapor intrusion science is, of course, not just *complex* because of the muddled nature of volatile organic compounds, which is largely attributed to the difficulty of accurately tracking and modeling the migration patterns and tendencies of these chemicals. The study of VI is also complex because scientists themselves are complex and challenged with navigating concrete worldly complexity: "It is better to think about the sciences as muddled rather than pure; to imagine the borders between the sciences and the worlds of language, culture, and politics as muddied rather than clear and distinct; to know scientists as complex hybrid figures rather than rarefied heroes; to see the work of the sciences as a complicated interaction with a messy world, an exchange involving tools, words, things, and even more nebulous entities, rather than a methodological, pristine encounter between mind and nature" (Fortun and Bernstein 1998:xiii). Vapor intrusion scientists, much like Endicott's plume residents, are caught up in real-world complexity and collectively struggling to understand.

The Regulatory Crux and a Plea for Vapor Intrusion Ethnography

The scientific complexity just discussed has of course complicated and augmented the challenge of developing national vapor intrusion policy.

Amid this scientific challenge is the growing fact that VI impacts many regulatory programs, including, but not limited to: Superfund Sites and Brownfields.[2] Some states have no vapor intrusion protocol or guidance for VI, and the Interstate Technology and Regulatory Council (ITRC) is the primary cohort of experts assisting states as they develop their vapor intrusion guidance and policies. A geologist with a consulting firm in Washington State recently said that the vapor intrusion pathway is already for some states and is quickly becoming for others "an inevitable regulatory issue . . . Understanding the risk posed by this pathway on property use and protection of human health will be a complex learning process as the regulators, environmental professionals, attorneys, real estate developers, insurance carriers and myriad of other stakeholders become better acquainted with this next *regulatory juggernaut*" (Brock 2009; emphasis mine).

While community involvement has been a grinding challenge for government environmental health agencies (see Little 2009), this crux was also a common theme to surface among the VI scientists and regulators I interviewed. During one of the plenary sessions at the EPA forum in Philadelphia in January 2009, David Polish, Community Involvement Coordinator for the USEPA's Region 3, discussed the community involvement challenges at vapor intrusion sites (Polish 2009). He contended that regulators and scientists working on vapor intrusion sites need to do a better job at community involvement and that the nature of vapor intrusion calls for new community involvement tactics. He explained that "in order for community relations to improve, scientists and regulators need to make five key changes to their practice." These changes, he argued, include "making an effort to meet with residents and community as soon as possible, and if possible, go to each home, business, school and 'sit down' with the people." He insisted that there was a need to "make a connection," "explain the vapor intrusion situation in common terms," and actively "share the data" generated by scientists. As I have argued elsewhere (Little 2013b), these are laudable signs of change in vapor intrusion policy that expose important inroads for both ethnography and political ecology theory.

Emerging VI debates are open to a plethora of critical anthropology questions. Recycling the same flood of questions employed in Kim Fortun's ongoing work on "exposure science" studies,[3] VI science is fertile territory for ethnography exploring questions including, but not limited to: how is transdisciplinarity practiced among vapor intrusion scientists; what is the array of scientific methods and imaginaries that come together in the work of vapor intrusion science; how are study designs for vapor intrusion science being developed, and how are different perspectives among scientists being worked out; what kinds of organizational structures support the work of vapor intrusion science, particularly its capacity for transdisciplinarity; how are vapor intrusion scientists reaching out to collaborators in other scientific fields; how are vapor intrusion scientists reaching out to policymakers, journalists, and community stakeholders in the public sphere; how is awareness of the need for and challenges of transdisciplinarity in vapor science shaping educational initiatives; what structural conditions have shaped the development of vapor intrusion science, and particularly its transdisciplinary dimension; how have technological advances shaped vapor intrusion science; what advances in other scientific fields have been critical to the development of vapor intrusion science; how have funding patterns and other economic determinants informed the development of vapor intrusion science; how have political trends shaped vapor intrusion science; and what key events, such as the IBM-Endicott spill, have shaped the identity of vapor intrusion science, among scientists, and among the lay public?

There are a number of sciences and cultures of expertise that dominate the emerging field of VI. ITRC has noted that VI introduces an interdisciplinary challenge because VI problems often require a combination of different people with different knowledge and expertise. Risk assessors, mechanical engineers, industrial hygienists, community relations coordinators, environmental scientists, soil scientists, hydrogeologists, analytical chemists, legal professionals, real estate agents, bankers, and insurance agents make up the regular whirlpool of individuals at any given VI site. One ITRC instructor at the training I attended in Portland, Oregon, in October 2008, pointed out that "VI is not only intrusive, but it forces you to work with people with different backgrounds. Often property value experts are more critical than having a soil scientist. Because it is not a simple pathway of exposure, VI is an interdisciplinary challenge." Nowhere in ITRC's VI two-day training is social science mentioned. I am, to my knowledge, the only social scientist pushing for more engagement with VI, which of course adds to the challenge. I am a subaltern in the sea of experts and stakeholders engaged in the contemporary VI debate. That crux is, of course, an exciting dimension of ethnographic practice.

Ethnographic research, from fieldwork to write-up, is a practice of twists and turns, an adventure of passages of clarity and many more encounters with confusion and ambiguity. The ethnographic research grounding this book has been directed and put to work to shed light on the complex human dimensions of high-tech pollution, deindustrialization, and the mitigation of an emerging scientific and regulatory problem: vapor intrusion. VI has been, until the writing of this book, a subject untouched by environmental anthropology. If fieldwork is, which I think it is, "the registering of sensory impressions in a (temporal) process of mutual subject-discovery and critique" (Borneman and Hammoudi 2009:19), ethnographic fieldwork *in* and *of* vapor intrusion debates surely helps open up the discourse and ensure that impacted residents' experiences with and perspectives on vapor intrusion get heard.

There are two reasons why I speak of both "in" and "of" in my formulation here. First, ethnography *in* VI debates is about applying ethnographic methods in the assessment of sites posing a VI threat. These debates are open to the disciplines and VI a porous environmental health conflict, and this porosity allows for inter- and transdisciplinary connection and description. This effort involves using a systematic approach to documenting and describing community responses to VI risk and VI mitigation. Ethnography, in this way, can be viewed as an interventionist practice revealing the multiple lines of evidence that make up the social environment conditioning the environment of vapor intrusion exposure and risk. And second, ethnography *of* vapor intrusion is about turning VI science and policy into a subject of study itself, its own "object domain" (Rabinow 2008) within the field of environmental anthropology and political ecology. An ethnography of VI calls for critical engagement with the anthropology of emerging sciences and technologies (Downey and Dumit 1997; Fischer 2003; see also Marcus 1995), not because this is the ultimate direction all anthropology needs to take, but simply because of the very fact that VI is emerging, under development, and so is the decision-making structure informing VI policy at the state and federal levels. In this sense, how the problem of vapor intrusion is being "composed" (Latour 2010b) matters.[4]

Aside from the flood of critical questions and openings for new research posed earlier, ethnography can assist VI policy and decision-making in a number of simple ways. As with its application in program development and interventions in social, public health, and environmental services, ethnography can be used to involve all stakeholders in providing input into VI decisions, thus promoting greater trust and ownership of the decision-making process and its possible outcome (e.g., mitigation). Renters are a growing population in the IBM-Endicott plume and their involvement or attendance at meetings is low, which calls on social science methods like ethnography to make these renters' voices and concerns audible and a part of the pool of "community stakeholders." Ethnography may also be a critical method to draw on to better describe and *monitor*—much like the groundwater monitoring wells marking the sidewalks in the IBM-Endicott plume—the process and dynamics of social change invoked by VI risk and mitigation. Embarking on this research would, of course, expose the difficult task of navigating and describing a complex confluence of intersecting subjectivities, environments, and epistemologies. My exposure to these networks is still in the making, as is my identity as both an anthropologist and a "stakeholder."

Being a "Stakeholder"

When I was invited to attend the EPA's National Forum on Vapor Intrusion in January 2009, I had no idea how my message, my interpretation of my research findings, would be received and interpreted. Here is a snapshot of what can be found in the EPA's proceedings document (EPA 2009:4) following the forum regarding my attempt to "represent" Endicott and communicate the import of social science in vapor intrusion debates, alongside other "community stakeholders," including Lenny Siegel with the Center for Public Environmental Oversight, Debra Hall who founded the Hopewell Junction Citizens For Clean Water grassroots organization, and Teddie Lopez from the Chillum TCE site in Maryland.

What follows is the interpretation of my presentation, which I titled "Vapor Intrusion and Social Science: The Case of TCE Contamination in Endicott, NY," as it appeared in the conference proceedings:

Mr. Peter Little described Endicott, NY, as a "poster child for VI and TCE remediation." Endicott was the location of IBM's first plant, which was sold in 2002, leaving behind a TCE contamination problem. There are currently 480 vapor mitigation systems in the community. Mr. Little, a Ph.D. student in applied anthropology, has studied the human dimensions and social impact of VI on the Endicott community, and the application of this information to policy and regulation.

Mr. Little described some emerging themes of his research, based on 35 completed interviews out of 60 to 100 planned. The community is not homogenous, he said; there is variation in community members' feelings about and involvement with the issue. Community members are concerned about health risk and property devaluation, and also that the community of Endicott is "dissolving" since IBM left the community.

The interviews revealed multiple lines of evidence of concern and frustration among community members. These included: contesting science and/or expertise; uncertainty (about the health risk and about the community's future); criticism of industry and government; and property devaluation. Mr. Little said that the tools of social science can be used to answer questions about issues such as local understanding of the social impact, the perspective of residents who do not take action, and the causes of community frustration and concern.

Mr. Little called the Forum a positive step in VI governance. Risk assessment and risk communication should be considered in the context of the actors, rules, mechanisms, and processes affecting the understanding of both risk analysis and how actions are taken. Mr. Little noted that public involvement is needed in the development of policy leading to pluralism and collective decision-making, as is a synthesis of the technical and social sciences.

There are a number of issues that could be discussed here, but the point about VI governance was especially interesting. In my presentation at the forum, I drew on the work of environmental sociologist Ortwin Renn, contending that his notion of "risk governance" (Renn 2008) ought to be adopted by government agencies, for it "looks at the complex web of actors, rules, conventions, processes and mechanisms concerned with how relevant risk information is collected, analyzed and communicated, and how management decisions are taken" (9). Aside from my explicit attempt to defend the integrity of social science at the forum, I did not offer much of an alternative "risk governance" solution, What follows is my vision for how VI mitigation policy and decision-making might be informed by knowledge drawn from monitoring practices that differ from those found in contemporary VI science spheres and studies.

Post-Mitigation Ethnographic Monitoring

Lenny Siegel, a vapor intrusion expert and co-founder and president of the Center for Public Environmental Oversight (CPEO), has been a technical advisor and advocate for communities nationwide facing VI risk. He has pointed out that "Depressurization systems are an effective form of vapor intrusion mitigation, and other technologies may be applicable as well. *However, they only work as long as they work.* To ensure that building occupants are protected, mitigation should be anchored in long-term management, which includes operation and maintenance, monitoring and inspection, contingency planning, notification, institutional controls, and periodic review" (Siegel 2009:8; emphasis in original). In addition to following up on VMS operations and maintenance (e.g., making sure the 90-watt fan works and the pressure gauge is working), Siegel has also argued, and I think rightly so, that

> those responsible for mitigation should periodically inspect slabs, seals, and other visible barriers. While in many cases this may be done infrequently, in some buildings—such as schools—it can be integrated into the daily routine of maintenance personnel—if they are properly trained. After initial tests show that depressurization systems are working, some agencies assume that installed systems continue to operate as designed. Others require periodic performance measures, such as sub-slab pressure tests. However, building occupants usually prefer indoor air testing, the best measure that the air is safe. The details may vary, but each site should be governed by a monitoring plan developed in consultation with building owners and occupants. (2009:8)

Post-mitigation monitoring of this kind may be a critical step in the right direction, as is actively engaging occupants of mitigated buildings in post-mitigation indoor air monitoring. The latter is perhaps most important, especially if a lasting goal of environmental health decision-making is to empower citizens by informing their understandings of their exposure and risk situation.

Federal and state regulators are generally opposed to continued testing of indoor air after mitigation, and this resistance reflects a neoliberal crux of mitigation in contaminated communities. Mitigation is considered a "good" decision, because it is "cost-effective." According to the U.S. EPA, installation costs for active venting systems like the ones populating the Endicott landscape range from $1,500 to $5,000, and the cost of annual operations and maintenance can range from $50 to $400 (USEPA 2008). Continued air monitoring and lab costs for analysis per home can range from $8,000 to $12,000 and can vary with monitoring duration. There is general agreement among environmental scientists and engineers that mitigation technologies do in fact do a good job of mitigating vapors, but to maintain the control of vapor intrusion these systems do require periodic maintenance and can't mitigate 100 percent of all volatized organics in the indoor air. As Peter Strauss with the CPEO contends:

> It is interesting that there is opposition to indoor air testing. Acknowledging that there are chemicals in a residence that can give false positives is only a reason to do a careful survey of those chemicals and remove them from the house. If this can't be done (e.g., perhaps the carpet cement is off-gassing), then the next best method is sub-slab soil gas sampling. This would involve drilling through a floor. It is interesting to note that EPA has done some research and it seems as though concentrations in soil gas drop off dramatically at some sites as you move away from the slab. The hypothesis is that gas may be collecting underneath the slab just waiting for someone to puncture the balloon (i.e., the slab) or that changes in barometric pressure allow soil gas to more readily vent to the atmosphere. Either way, in my opinion, sub slab or indoor testing is preferred over estimations from how far you are from a plume, what the attenuation factor is, etc. This is not to say that every home in America needs to have its air tested, but those close enough to a source that has probable chemical mobilization should go through a step by step process . . . to determine whether it should be tested. (Strauss 2010)

As discussed previously, vapor intrusion science and regulation is currently under construction, and inroads for the anthropology of this emerging problem can be forged. A major step toward participating in VI discourse and decision-making will require collaborating with citizens and scientists engaged in VI debates. As an engaged anthropologist and political ecologist, there are a number of reasons why I think this effort is both critical and timely.

First, employing ethnography is one way to provide an "evidence-base" for the evaluation of alternative research projects that could result in beneficial VI institutional and research program changes. One such change, as I have argued in this chapter, is a shift toward community-based air monitoring in communities that have been mitigated for VI. This effort, of course, is based on the assumption that to carry out such an alternative VI research project might require experimenting with and learning *how* to position oneself inside the emergent VI controversy,[5] a crux made up of the twin challenge of making anthropology and political ecology theory matter in the science and policy world of VI (Little 2013b).

As a problem-oriented ethnography of technological disaster mitigation, the research presented in this book, despite the always built-in limitations, can inform the way we think about vapor intrusion and mitigation and can open up opportunities for new discussions and even new science futures and vapor intrusion policies. Uncertainty, in this sense, is not necessarily a downfall of the current situation. In many ways, because VI debates are elusive and saturated with ambiguity, they call for better public engagement and citizen-science collaboration, especially because, as it has been argued, "Public demands for information and involvement in decision making are often loudest where there is most conflict and where ambiguity, both political and scientific, is greatest" (Tait and Bruce 2004:413). VI policy is in its infancy and citizen involvement, as the Interstate Technology and Regulatory Council (ITRC) team of experts explain, has played a small role in the decision-making process. VI experts talk about how each VI site is unique, yet this said diversity is not ethnographically investigated. In other words, "If we simply assume, rather than empirically identify, those dimensions of community most affected by environmental risks, can mitigation strategies be successful" (Omohundro 2004:21)? Furthermore, as an emerging environmental health science, we need to listen to experts in the field to better understand how they personally envision their own role as scientists and experts working in communities disturbed by vapor intrusion. In an interview with a NYSDEC official working on the IBM-Endicott site, he shared with me his ideas about finding a way to help residents at VI sites inform their own understanding of risk and therefore provide them with the evidence—based on citizens' own household indoor air sampling—to inform their own perspective on the efficacy of mitigation decisions:

> I would like to distribute passive sampling devices, like the ones that are used for radon, so the homeowner can have them in the basement for 30 days and then have them send it back to the lab and then we would know if there is contamination there or not. It would give them a little bit more control over what happens in terms of mitigating or not mitigating. That is where, personally, I would like to see us go in the next five years or so. I think right now, you know, people get a letter from us or from the health department that says you are potentially being exposed and we want to mitigate your house and it puts people in a position of being concerned about their health and about their property values.

While handing out sampling devices to residents so they can "see for themselves" how mitigation technology works might result in residents becoming more "scientific" about their sense and understanding of risk, this approach misses a critical point that is at the heart of this book: mitigation is indeed *more* than the diversion of toxic vapors when it deals with human beings managing to live comfortably in space of coupled contamination and deindustrialization. Mitigating IBM's birthplace, in this sense, involves people, not just intruding toxins; volatile people, not just volatile organic compounds. That is a difference that really matters. It is about going beyond risk, beyond the deployment of high-tech risk detection devices,[6] beyond mitigation, by looking at and taking seriously how both are in fact "experienced, narrated, and known" (Reno 2011: 527).

This all brings into bold relief the political ecology of emerging vapor intrusion science and policy, which is to say that re-thinking vapor intrusion science and policy will require the corporate state to re-adjust or reshuffle their "cognitive equipment" (Latour 2007) and embrace a more concrete angle on accountability and diversify modes of representation: "To be accountable . . . is exactly this: To be able to give an account, and to be made responsible for what you conclude from it. Without calculative devices, politics is emptied; limited to calculations, politics is gutted" (Latour 2007:25). Of course, how "accounts" are made always calls attention to certain politics of representation. Citizens and experts engaged in emerging vapor intrusion problems will forever confront the cruxes of "coming together": "In one way or another, the question of representation is central and can only be asked if those who come together accept the risks and challenges to which that coming together obligates them" (Stengers 2011:394).

8

Accounting for the Paradox of IBM's "Smarter Planet"

We are taught that corporations have a soul, which is the most terrifying news in the world.
—Deleuze (1992:6)

We must never feel satisfied.
—IBM's founder, Thomas J. Watson, Sr.

Ultimately ethnographic fieldwork "is an educational experience all around. What is difficult is to decide what has been learned" (Geertz 2000 [1968]:37). It is easy to say that I have learned many things from spending time talking to residents of Endicott's IBM plume. Deciding what to write about, what to cultivate, what to expose and attend to was no doubt a challenge. The journey from fieldwork to write-up is a winding road with unexpected curves and occasional dense fog, yet several ethnographic questions served as my guide on this twisting and uneven terrain: how do Endicott residents understand and talk about IBM's toxic plume; how do residents living in "mitigated" homes situated over the IBM toxic plume understand and talk about vapor intrusion risk and the mitigation effort; and what prompts citizens to take action and become vocal community stakeholders, and how does or doesn't environmental justice figure in grassroots action in Endicott?

Certainly, the enduring ambiguities of the TCE plume lurk and the plume's extralinguistic material reality persists. But, as this book has aimed to show, residents in the plume talk about interconnected things, they vocalize meshed matters of concern, reminding us that local discourse and especially "the salience of personal experience" (Farmer 1999:xli) matters. After numerous interviews and conversations with Endicott's plume residents, it became clear to me that residents spoke freely about their intersecting experiences with IBM deindustrialization and the consequences of both TCE contamination and mitigation. The activists I spoke with openly discussed their advocacy and shared with me the personal experience of being a vocal community stakeholder. Every interview experience helped me think through the tangle of socio-environmental problems in focus.

In many ways, this book has honored the fact that the struggles and troubles unfolding in IBM's birthplace are existing *problems* that call for critical diagnoses and reflection. I tried, as Foucault would put it, to "analyze the process of 'problematization'—which means: how and why certain things (behavior, phenomena, processes) became a *problem* . . . I have tried to show that it was precisely some real existent in the world which was the target of social regulation at a given moment" (Foucault 2001 [1983]:171). Vapor intrusion was that "real existent," that target of regulation and intervention that led to a pre-emptive mitigation decision in the IBM-Endicott plume. Local discourses on vapor intrusion risk and critiques of the efficacy of mitigation became a strong focus of my research as I learned that these problems were a real source of agitation and frustration, that they were problems meshed with and shaped by other problems of late industrialism (Fortun 2012), namely deindustrialization, factory corrosion, techno-scientific authority, and corporate-state power. Living in the plume remains elusive, even as IBM and the state claim they have adequately addressed the TCE vapor intrusion threat. The book, in this sense, honored a certain kind of empiricism:

> An empiricism that admits that one never gets to the bottom of things, yet also accepts and even celebrates the disavowals required of us given a world that forces us to act. An empiricism that is ethical because its methods create obligations, obligations that compel those who seek

knowledge to put themselves on the line by making truth claims that
they know will intervene within the settings and among the people they
describe. (Rutherford 2012:465)

Ethnographic practice taught me that the mitigation of TCE contamination has not necessarily mitigated plume residents' sense of environmental health risk, nor softened frustrations over IBM deindustrialization. In chapter 3, I looked at how intersubjective experience informs
plume residents' understandings of TCE risk, highlighting local environmental health politics of TCE vapor intrusion. I engaged tough
questions: What is the experience of residents living with/in the IBM-
Endicott plume? What do vapor intrusion scientists and New York State
officials say about the risks residents face? How do Endicott's plume
residents tangle economy and environment, doubt and distrust as they
talk about contamination and community change? What remains elusive amid aggressive mitigation efforts? What do people's understandings of risk and mitigation efforts tell us about what they know? Guided
by these questions, the chapter elucidated plume residents' experiences
with and perceptions of TCE risk and IBM deindustrialization and its
entangled side effects. It aimed to document what I see as the "one-two
punch"—IBM deindustrialization and TCE contamination—experienced by many of Endicott's citizens. Paralleling this "one-two punch"
scenario is an effort to soften the blow of TCE contamination with
state-led technocratic solutions. The remediation of the IBM-Endicott
plume and the mitigation of TCE risk has been the focus of a long-term
multi-government agency effort.

How this mitigation effort is understood and experienced by the
"mitigated" was the focus of chapter 5. I showed that these "mitigated"
residents don't feel that IBM has fixed the TCE plume. They contest
IBM's claim that vapor mitigation systems were a "gift" to residents and
that mitigation is a "Magic Bullet" remedy. This was a strong "fact" that
I hope the book communicated, and this finding has also helped me
better understand the need to afford vapor intrusion risk and mitigation some degree of subjectivity. Ethnographic work attends to this subjectivity and takes lived experience and narrative seriously, especially
when people are discussing serious matters, like TCE contamination
and living in a TCE plume.

Chapter 6 explored the emergence and influence of grassroots action amid this contamination conflict. When I returned to Endicott in the summer of 2008, I wanted to learn more about the members of the Western Broome Environmental Stakeholders Coalition (WBESC). I wanted to know (1) what inspired residents to take action and get involved in the IBM-Endicott contamination debate, and (2) to what extent these activists framed their struggle in terms of environmental justice. The second focus was critical to my research because I was curious to know how WBESC activists reacted to the National Environmental Justice For All tour, which stopped off in Endicott in October 2006. I found that these activists were somewhat perplexed when asked to talk about the tour and the idea of environmental justice. Some insisted that environmental justice had nothing to do with the contamination conflict, while others felt it did because there was a growing minority population and a surging welfare population that was changing the demographics and the "feel," as one resident put it, of the community. This chapter highlighted the ways in which environmental justice politics are locally, and it aimed to illustrate the micropolitical dynamics of the environmental justice movement in general and alliance-building politics in particular.

In chapter 7, I discussed the ways in which I see anthropology contributing to and informing vapor intrusion policy and decision-making. I also engaged my own attempts to be an advocate and "community stakeholder." The real active ingredient of the chapter was to think about anthropology *in* vapor intrusion debates and its application in a field dominated by engineering and environmental science. Throughout this effort, I remained cognizant of a very real challenge for anthropologies of environmental public health problems, which is that these problems are indeed complex and ambiguous. More often than not, environmental health risks are "trans-scientific" (Weinberg 1972), problems beyond the capacity of science to fully answer, problems plagued by "undone science" (Hess 2007a), and "research silences" (Brown 2007:261).[1] Plume residents understand that cancer is a complex illness with a complex etiology. They understand the force of uncertainty. Experts with the ATSDR and the NYSDOH understand that cancer is a multifactorial disease, yet from an epidemiological standpoint, it, like all other diseases, does not occur randomly. Both impacted residents and epidemiologists agree that cancer is not random, begging insight into challenging questions: why do I have non-Hodgkin's lymphoma? Why does my neighborhood seem to have a lot of cancer? A health statistics review or even a well-funded epidemiological study will likely determine that the cause-effect relation is hard to pin down, that all studies have built-in bias, and that all scientific studies can be re-studied with a different research design and possibly find different results.

This is the confluence of scenarios and points of contingency that contribute to both ethnography and environmental health science. In the end, both cope with comparable epistemic handicaps (e.g., dose-response relations and culture-intentionality relations). Whitehead (1978) was right to say that both social and natural scientists share the goal of intelligibility. We "frame a coherent, logical, necessary system of general ideas in terms of which every element of our experience can be interpreted" (3). I think this is a good way to think about the hybridization of anthropology and vapor intrusion science. Plume residents live-with-the-plume and live-with-mitigation and they live the ambiguity that attracts and fuels scientific interests. To inform vapor mitigation decision-making and make these risk governance practices more sensitive to community concerns and lived experience, and more open to ethnography, I proposed a post-mitigation approach that would more meaningfully involve "mitigated" residents in indoor air sampling to help them determine the efficacy of mitigation.

In addition to being a problem-oriented ethnography and political ecology study, this book has ultimately been anchored by a strong "local" conflict focus, despite moments of "multi-sited" connection and conversation with scientists and regulators at the state and federal levels. While the study was guided by a political ecological theory of human-environmental relations favoring micro-macro processes and contextualization, as well as the vibrant interconnections between people, place, capital, agencies, expertise, science, technology, and power, it has also been an ethnographic account committed to a "situated" people of a contaminated community coping with elusive mitigation. This is, I would argue, both its strength and its weakness. Regarding the latter, it was not comparative in scope, yet this angle or approach could surely inform the direction of future environmental anthropologies of contamination in general and risk mitigation in particular. Vapor intrusion risk is in fact *distributed* across space and is a growing concern in communities struggling with harmful contaminants like TCE, a reality that calls for more, not less, critical ethnographic research and insight.

The IBM-Endicott toxic struggle exposes no shortage of problems for ethnographic inspection, no shortage of discourses and practices to attend to and pay attention to. Carrier is right when he says that "[o]ur attention to local contexts . . . needs to be complemented by an awareness of their relationship with larger contexts, and hence indirectly with other local contexts" (2004:14). It might also be argued that our attention to toxic struggle ought to be open to theoretical commingling and the relationship between theory development and locality. That said, there was a two-pronged ethnographic ethics in circulation throughout the research and writing process: (1) to listen to residents of the IBM-Endicott Superfund site and make their understandings audible and their struggle describable, and (2) to draw on these discourses to rethink toxic struggle, corporate power and responsibility, and develop a political ecology for navigating the emergent *mitigation landscape*.

Revisiting the Political Ecology–STS Synthesis

Scholarship within political ecology is important for understanding human-environment interactions, science-citizen interactions, and community-institutional relations. Political ecology, as a general theory of environmental politics or ecopolitical ontology, has been especially attractive for researchers wishing to articulate the ways in which citizens and communities engage with (and often resist) state organization and control of both the environment and the economy, both ecology and capital. I employed the *mitigation landscape* concept throughout because I felt that in order to understand and honor the human-environment interactions and points of friction in this historic high-tech company town, my political ecology perspective needed to open up to scholarship critically addressing the concrete social and political dynamics of risk, neoliberalism, environmental justice, and technological disaster. This synthetic political ecology was both a point of departure and framework to fall back on when I found myself struggling with, alongside Endicott residents, the problems of "toxic uncertainty" (Auyero and Swistun 2009), "powerless science" (Jas and Boudia 2011), and the burdens of epistemic vacuity so common in contaminated communities.

As argued in this book, a political ecology perspective nestled at the interface of anthropology and STS exposes the volatile nature of vapor intrusion risk debates and the unfinished and ongoing techno-scientific politics of risk mitigation. While listening to the experience of Endicott's plume residents was a central motivation of this book, studying the contentious mitigation of vapor intrusion risk is only useful if it takes seriously what scientists say and how they act ethically and in "coded" ways:

> Anthropological works that engage the sciences as domains in which ethics are worked out . . . assume that the sciences and scientists are always on the move, changing in response to shifting contexts, always altering perspectives on what is real, natural, inevitable, possible, and obligatory. Scientists must forever read the landscapes within which they operate; their imaginaries are necessarily historically inflected and socioculturally sedimented. Like other cultural producers, they operate within ethical plateaus that forcefully mold how they think and talk about their scientific work and about the wider world in which their science is supposed to function. Scientists are, inevitably, coded—by the technologies with which they work, by hegemonic cultural formations, and by forceful political-economic currents" (Fortun and Fortun 2005:50).

The present work tuned in to the complex social production of risk mitigation and the vibrant "epistemic culture" (Knorr-Cetina 1999) and challenges faced in the production of contemporary vapor intrusion science and policy. State scientists and regulators know that mitigation is one step in addressing vapor intrusion risk. Mitigation is "for the most part, straightforward," as one NYSDEC official told me. I sensed from discussions with VI experts that the mitigation of VI is not plagued with the same "knowledge gaps" (Frickel and Vincent 2011; see also Hess 2007a and Latour 1987) that VI risk assessment is. I also observed that these same scientists and regulators spend little time focusing on how mitigation is concretely experienced by residents living in homes declared "mitigated."

This book attended to this knowledge gap and along the way aimed to show how mitigation is not uncoupled from social reality, not an inert object of risk control. Mitigation, as Bruno Latour would have it, is an actant, just like trash, weather, stem cells, or food. Mitigation is not a background thing on the IBM-Endicott plume landscape; it is experienced by real people living in an *uncertain* environment of exposure and *certain* environment of deindustrialization and corporate deterritoriality.

The political ecology strategy explored throughout, aimed to expose the edges and blurred boundaries between IBM deindustrialization, contamination, and mitigation, and how these processes and strategies generate life politics and shape political ecology struggles.[2] In general, it could be argued that most political ecology critically navigates the paradoxical play of economic "gains" and socio-environmental "losses," trying to better understand how these zones of interaction—where economy, environment, science, and the state are tangled—are humanly experienced. This book has aimed to contribute to this thread of environmental anthropology theory and debate by strengthening the knot of political ecology and STS by focusing on the human dimensions of the reasoned game of toxics mitigation. The *mitigation landscape* concept played a central role in bringing these theories and strategies together and it was deployed to experiment with and rethink debates over "risk society" (Beck 1986, 1998; Giddens 1990, 1991) by attending to the mitigatory ironies of high-tech modernization, deindustrialization, and contamination.

The term probed certain dynamics of risk politics emerging in a community context where risk mitigation strategies have in turn reinforced the power of IBM, the state, and VI science and expert authority. Risk society, or what has been referred to as "incomplete modernity" (Rustin 1994), is based on the notion that modern institutions (e.g., government agencies and corporations) are characterized as having no settled futures, so they appear mostly as science- and technology-based societies heavily invested in risk management. Global warming, toxic substance intrusion, and global economic recessions are all reflections of risk society processes. They also happen to be risks that increase the power of science and expertise at the same time that these risk debates illuminate the ironies of science, as the politics of uncertainty and "not knowing" are enduring dominant themes of global risk society.

In such a world of risks, the question of "What if?" is amplified, even becoming an essential influence on individual and social life. The parallels to the IBM-Endicott technological disaster are front-and-center:

What if TCE exposure has caused increased cancers in the plume? What if vapor intrusion and mitigation are not the exact science experts claim it to be? What if more of us sought to better understand what "scientists themselves understand as worthy of care and ethical attention" (Fortun and Fortun 2005:44)? What if we viewed the IBM-Endicott disaster as a cocktail of interconnected matters of fact and concern (Latour 2004), a milieu of incorrigible dispute? These have been the lurking questions guiding the present work.

A Landscape of Litigation Drag and Irresponsible Smartness

In addition to the understandable frustrations Endicott residents face when confronted with IBM deindustrialization and contamination, many are simply agitated by the fact that IBM is fighting residents in court. Litigation was a real source of distress for John, a retired IBMer who has lived in Endicott since the early 1960s: "By the time this is settled and if this is even going to be settled, we will all be dead. This whole lawsuit will save IBM millions. They have more money and lawyers than you can believe." There is a sense among the plaintiffs I interviewed that nothing can or will restore what was lost in Endicott as a result of IBM deindustrialization and contamination. This experience of litigation stress and distrust is common in contaminated communities and zones of disaster. As Button points out, delay is a powerful tactic of corporations responsible for disasters, for it is a "testament to the enormous resources of the world's most profitable corporation to prolong litigation and wear down plaintiffs and their financial and legal resources" (Button 2010:171). It does not take a stretch of the imagination to wonder why residents living in the IBM-Endicott plume are worn out: IBM is a multi-billion-dollar, multinational corporation. In 2011, the company's net profits neared $107 billion. The power asymmetry is self-evident and the "burden of proof" placed on residents to prove toxic harm has been done is a chronic struggle experienced by plaintiffs engaged in class action trials in contaminated communities (Michaels 2008). Moreover, the enduring problem of corporate denialism generates further stress for residents and plaintiffs of contaminated communities, as corporations like IBM claim, again and again, that they have done no harm (Markowitz and Rosner 2002).

But the litigation distrust I witnessed was as much about power asymmetries and public health concerns as it was about IBM deindustrialization, IBM abandonment, and hurtful corporate deterritoriality. Many plaintiffs are especially frustrated because IBM is defending itself against residents who spent their working lives in the IBM plant assembling electronics and engaging in research and development to help make IBM a Fortune 500 company, a true "business machine" of capital accumulation and corporate leadership. Many feel they are being dragged into a narrative of dystopia—perhaps even "neoliberal creative destruction" (Harvey 2007)—and that the TCE plume isn't going away anytime soon and the class action lawsuit will never actually result in meaningful compensation. Plume residents are stuck with a situation of "difficult forgiveness" (Ricoeur 2004), despite the efforts of IBM and the state to clean up and mitigate a toxic mess. "If forgiveness is difficult to give and receive, it is just as difficult to conceive of" (ibid.:457). This rings especially true in a deindustrialized and contaminated community like Endicott, where IBM began, where IBM busted, and where IBM left behind a "clash of community" (Nash 1989) and risk situation that is not easily, if ever, forgivable or corrigible. The situation, in other words, is a powerful techno-economic and environmental disaster that is "irreversible" (Callon 1991).

As this book has attempted to show, the act of mitigation is, in its most powerful symbolic form, an *acting* ghost of IBM reterritorialization, and I use the word *acting* here with IBM's own interest in the term in mind: "Acting—the last step in the difficult process of making the world work better—should really be a formality" (Maney, Hamm, and O'Brien 2011:310). The "civil corporation" (Zadek 2001), if there is indeed such a thing, is after all "one that takes full advantage of opportunities for learning and action in building social and environmental objectives into its core business by effectively developing its internal values and competencies" (ibid.:37). The struggles exposed in this book are themselves "opportunities for learning and action." Slowing down and paying attention to people struggling with IBM pollution—a serious test of patience for high-tech firms addicted to ever faster processing—should also be a formality or learning goal of IBM, especially if they wish to build a "civil" relationship with communities it has contaminated and mitigated for harmful toxic substances. These new relations cannot be rightfully based on the fetish of nanoseconds and high-speed computing, but instead on slow, long-term commitment and active interest in and solidarity with people impacted by IBM production and socio-environmental plunder. IBM's new president and CEO Virginia M. Rometty, who replaced Samuel J. Palmisano in October 2011, ought to look back at a few lines from the chorus of IBM's famed company song, "Ever Onward": "We can't fail, for all can see that to serve humanity has been our aim."

High-tech and high-speed life is not marked by innovations in our capacity to listen to people, to sit down with them, to feel empathy for others, to slow down. These forms of interaction and intersubjective practice call for a slower pace of life, perhaps slower ways of conjuring and understanding the world around us. Real IBM corporate social responsibility is a form of responsibility that deals with talking, listening, even sitting down with people in the IBM-Endicott plume, not "settling" with them in the court. The latter seems more irresponsible, less "smart."

I am left, like many residents of the IBM-Endicott plume, with enduring questions and durable ambiguities, always confronting the problem of knowledge "gaps" and "holes" (Burt 1992; Frickel and Vincent 2011; Hess 2007a; Latour 1987) and contemplating the "deceptive screen of expert assurance" (Žižek 2010).[3] The plume that "lurks," as one resident phrased it, remains mysterious no matter how many studies or surveys are fielded, no matter how many books on the subject are written. It is a matter of concern that is vibrant and rich with uncertainty, as is the recent "hydrofracking" debate dominating environmental politics in this region of New York in the last decade (Wilber 2012). This has become a serious environmental conflict swarming around Endicott and neighboring towns, because the Marcellus shale of the region is a "hot spot" for natural gas drilling and the even more contentious issue of intrusive drilling techniques like hydraulic fracturing (or hydrofracking) which releases natural toxins embedded in the shale that can be carcinogenic. Also, toxins from fracking additives, like alcohols, petroleum distillates, aromatic hydrocarbons, microbiocides, and glycol ethers, which is a common ingredient in antifreeze, pose human health risks. Amid this emerging battle, the politics and uncertainties of the mitigation landscape explored in this book become even more amplified. The contemporary landscape and political ecology of this region of New York is now, more than ever, one of interconnected forms of environment disaster and risk. Both the threat of vapor intrusion and the environmental ruin experienced in the region as a result of aggressive shale gas exploration has only amplified and enriched the politics of risk unfolding in this transformative, uncertain, and distressed region of upstate New York.

Perhaps we ought to keep in mind, as Amartya Sen points out, that a "rich phenomenon with inherent ambiguities calls for a characterization that preserves those shady edges, rather than being drowned in the pretense that there is a formulaic and sharp delineation waiting to be unearthed that will exactly separate out all the sheep from all the goats" (Sen 2003:xiv). It is my hope that this book has reflected such a characterization and has provided an enriched ethnographic account of the lived experience of Endicott's residents who struggle with the ongoing vortex of contamination and their environment of precarious mitigation.

If IBM is in fact interested in "building a smarter planet," the planet it is referring to needs to be radically re-conceptualized, as does "responsible" corporate response to technological disaster. Is what has unfolded in Endicott, "IBM's Birthplace," an example of an IBM acting socially and environmentally responsible? It is, after all, "IBM's policy to conduct itself ethically and lawfully in all matters and to maintain IBM's high standards of business integrity."[4] When IBM "formalized" environmental protection into its corporate policy in 1971, that move involved a "commitment to leadership across environmental areas ranging from energy efficiency and water conservation to pollution prevention and product stewardship."[5] What happens when such a "commitment" falls short of being "responsible"? What happens when corporate community damage reaches a point of irreversibility, immitigability, and incorrigibility?

The Endicott case exposes the fog of responsibility. Under closer inspection, as this book has shown, responsibility is not *just* about IBM being a more socially or environmentally responsible corporation or state agencies doing a better job of serving citizens by doing better hazardous waste cleanup. Rather, acting responsibly is instead about the corporate-state accounting for the vibrant contradictions at work both in spaces of "responsible" pre-emptive mitigation decision-making and where hurtful litigation decisions have been made. Is not the toxic struggle unfolding in Endicott a time of opportunity, a time for IBM to turn to its own ethos of innovation to re-THINK its own ethical and lawful actions? THINK is not just a clever business slogan, but a critical concept to put to work and use to reflect on the lives, understandings, and concepts explored in this book.

For IBM and the state to seriously re-THINK their actions at the Endicott plume site, they must thoroughly confront *thinking* as a conceptual, political, and ethical practice. Being "responsible" actors calls for recognizing, even *admitting*, what really matters most in a "contaminated world" (Omohundro 2004), in a world of disasters, both large and small, natural and technological, which is to "admit that the environment is actually inside human bodies and minds, and then proceed politically, technologically, scientifically, in everyday life, with careful forbearance, as you might with unruly relatives to whom you are inextricably bound and with whom you will engage over a lifetime, like it or not" (Bennett 2010:116). IBM and the State, in this sense, are better off giving up "the futile attempt to disentangle" (ibid.) the contaminated landscape from the deindustrialized and mitigated landscape. Why not welcome and learn from the "trenchant materiality" (111) of community experience with corporate deindustrialization, contamination, and mitigation, as well as the incisive moral paradoxes of corporate responsibility?[6]

It is time to re-THINK the "cognitive equipment" (Latour 2007) of states and corporations, especially powerful corporations claiming that "the signs of a smarter planet are all around us" and that this smartness is realized when "We've learned that our companies, our cities and our world are complex systems—indeed, systems of systems. Advancing these systems to be more instrumented, intelligent and interconnected requires a profound shift in management and governance toward far more collaborative approaches."[7] Furthermore, the "Smarter Planet" agenda is for IBM more than crafty word play: "This idea isn't a metaphor, or a vision, or a proposal—it's a rapidly emerging reality."[8] A profound *shift* is needed indeed, but perhaps not the *shift* that IBM (or the State) has in mind. The so-called Decade of Smart may in fact require a shift toward greater ethical experimentation, greater attention to birthplace honor, greater attention to the complex political ecology of high-tech pollution and emerging paradoxes of "responsible" risk mitigation. It calls for a shift in thinking that welcomes the notion that "All thinking demands a *stop*-and-think" (Arendt 1981:78; emphasis in original). This book has been a "stop-and-think" exercise, a stop-and-think diagnosis of interconnected social, economic, political, health, and environmental troubles in IBM's birthplace community.

As one resident told me, the simplest and best thing IBM could do is to admit that wrong was done, even if vapor intrusion mitigation is the right thing to do to actively reduce TCE exposures. In his words: "For many years, IBM was the Messiah of this region. There are many people, including our former mayor, who still strongly believe that IBM could do no wrong and have done no wrong. I just believe IBM is draggin' its feet. It has caused a loss of faith in the community. They have made token gestures to give the impression that they care, but they really messed up here." It is with these words of reflection that I am compelled to continue asking myself when thinking about Endicott's enduring toxic struggle: is mitigation a token gesture of IBM caring, of IBM being smart, of IBM building a smarter Endicott?

As noted by IBM's former president, chairman, and chief executive officer, Samuel J. Palmisano, a Smarter Planet is one that goes beyond the challenges of building bigger and better "technology": "the barrier is no longer technology. What we make of this new reality will depend, rather, on how we pursue the timeless goals of all social and economic systems—reliability, trust, fairness, inclusion, sustainability, human rights, prosperity and individual empowerment. I believe we must do so in very new ways. Building a smarter planet isn't simply a recipe for economic growth; it's also a recipe for radically expanded economic and societal opportunity. It's not just a way to make the planet more efficient, but also to make it more sustainable. It's not just a way to do well by doing good; it's also a way to do good by helping all the world's regions and people do well."[9] Welcoming "new ways" of thinking and doing for an old company with deep corporate cultural roots is no doubt challenging. Even more challenging, it seems, is to welcome and mesh political ecology thought with corporate interests in building planetary-scale efficiency goals that aim to help humans "do well." While IBM turns to making the planet smarter—an audacious mission to say the least—it ought to take the opportunity to slow down, revisit its birthplace, and listen to how people are doing there to learn something about the real socio-environmental life struggles and lifetime exposures to the tangle of corporate deindustrialization, contamination, and risk mitigation. IBM should re-THINK its approach to socio-environmental caring and reflect seriously on its own practices of birthplace destruction and mode of "staying with the trouble" (Haraway 2011) IBM helped create. IBM should not, of course, only be held "accountable" after toxic plunder and during mitigation, but all techno-capital firms ought to find alternatives in the planning and production process to minimize and hopefully eliminate future toxic spills and the amoral production of late industrial "sacrifice zones" (Lerner 2010).

IBM's THINK motto is actually useful to think with in closing, as it is an idea and process that needs to be rediscovered and repurposed. We shouldn't treat THINK as a bare corporate tagline, as syntactic residue of the anchoring philosophy of IBM's Thomas J. Watson, Sr. I often find myself thinking about Endicott's mitigation landscape, revisiting mitigation in new ways each time it comes to mind, but one critical thought sticks: the act of mitigating the wrath, pain, grief, and displeasure of contamination is more difficult than IBM and the state are making it out to be. Mitigation is as messy and precarious as contamination and deindustrialization. Residents are not *exposed* to one or the other since these processes are enmeshed. As experienced by residents, homes, as dwelling spaces and places, are not contaminated one day and mitigated the next. The plume isn't contaminated one day and mitigated the next. On the ground, among plume residents, that bifurcation is confusing, as is the splitting of the troubles of corporate deindustrialization from the struggles of contamination and uncertain mitigation. All these clean breaks are examples of boundary logics imposed by IBM and the state.

This book was informed by nomadic ethnographic practice and an "adventure of ideas" (Whitehead 1933).[10] Along the way I experienced ethnographic, theoretical, and moral twists and turns, with the greatest crux being how to think for myself while honoring the concerns and life politics of Endicott residents. Navigating and choosing strategies of articulation was not always easy, but tuning in and listening to residents' life experiences was. Ethnographic practice has taught me that people—residents, scientists, government officials, industry representatives, etc.—really do matter *in* toxic spills and *on* what I have called the *mitigation landscape*. Endicott, as a corporate "sacrifice zone" (Lerner 2010), will forever be a landscape of transformation, of ontological quandary, techno-capital neglect, of meshed life struggles, an ecopolitical landscape *par excellence*.

A certain anthropology of neoliberal space informed the political ecology explored in this book, as the *mitigation landscape* was seen as a space of bodies, knowledges, economies, buildings, toxins, and mitigation technologies "materially intermingled" (Murphy 2006:169). Our focus was a new "appropriated space" (Serres 2011) marked by the spatial expansion of vapor intrusion risk and technologies of toxic vapor diversion. It was a study of yet another corporate-state mastering of space with the tools of techno-science and an administrative ethos of risk management. It was a political ecology of high-tech disaster accounting for multiple forms of disaster and distress which have transformed life and living in a New York community in a neoliberal age.

I tried to show that Endicott is a place and space shaped by an "ecology of practices" (Stengers 2011) and a community where "smarter" moral action can take place. While "the lexicon of calamity is . . . deeply colored with moral hues" (De Waal 2008:xiii) and is always in-the-making and being re-composed in zones of disaster, surely a "smarter" political ecology and anthropology of high-tech disaster is one marked by greater exposure of neoliberal environmental paradox. We live in a world of many hollow corporate taglines, responsibility strategies, and planetary imaginaries that too often overlook, dwarf, or even seek to discredit the concerns and frustrations of people living in zones of industrial disaster.

Alfred North Whitehead once said that "We think in generalities, but we live in the details."[11] Building a "Smarter Planet," in the end, requires smarter ethical experimentation, smarter cognitive and behavioral reflection, and smarter THINKing about and learning about the fine details of socio-ecological life and recognizing that "[w]e learn to think by giving our mind to what there is to think about" (Heidegger 1968:4). This thinking and learning involves embracing the intrusive tangle of ecologies of disaster, uncertainty, and power that citizens struggle to navigate and live with. Accounting for and cognizing this knotted reality matters, especially if the planet IBM continues to refer to is understood as a planet anthropomorphized now more than ever before. Returning to and attending ethnographically to troubled *birthplaces* of deindustrialization, contamination, and risk mitigation can help us learn how people's entanglements with science, technology, corporate-state power, and politics are actually experienced, what the political ecology of a Smarter Planet might look like. A "smarter" outlook or vision, I dare say, is one that pleads and thirsts for more moral ecological adventure and ecopolitical description, not less. We need more recognition of emergent ecologies and ecological complexity, not less. We need more corporate and state accountability and caring, not less. Our times call for more acknowledgment of and moral reflection on the lived experience of toxic struggle, community corrosion, and the various cocktails of trouble and distress invoked by technological disaster, not less. We also need more critical diagnoses of the political ecology of high-tech capitalism and ultimately more THINKing about new mitigation landscapes, not less. We need all of this and more to help us size up and interrogate the terms and planetary ethos of IBM's Smarter Planet mission. What better place than a birthplace to explore such an alternative discussion and course of action?

NOTES

NOTES TO THE PREFACE

1. As Bruno Latour notes, "Digital democracy has generated a lot of hype, but . . . its true development is still to come and that it will be necessary to invest also, in no small part, in the theoretical import of the notion of network" (Latour 2010a:17) since the "expansion of digitality has enormously increased the *material* dimension of networks: the more digital, the *less virtual and the more material* a given activity becomes" (2010a:8, emphasis in original).

2. Retrieved at http://www.ibm.com/ibm/responsibility/awards_recognition.shtml, May 27, 2012.

3. Retrieved at http://www.ibm.com/smarterplanet. Accessed May 11, 2012.

4. Retrieved at http://www.ibm.com/smarterplanet/us/en/events/sustainable_development/12jan2010/html/index.html. Accessed May 11, 2012.

5. In July 2012, for example, IBM signed a collaboration agreement with the Tanzanian Ministry of Communication, Science and Technology to "help accelerate the adoption of technology as part of Tanzania's ongoing development and strategy to increase its competitiveness in East Africa." Retrieved at http://www-03.ibm.com/press/us/en/pressrelease/38243.wss, July 6, 2012.

6. Retrieved at http://www.ibm.com/smarterplanet. Accessed May 11, 2012.

NOTES TO CHAPTER 1

1. Superfund is the name given to the environmental program established to address abandoned hazardous waste sites. It is also the name of the fund established by the Comprehensive Environmental Response, Compensation and Liability Act of 1980. This law was enacted in response to the discovery of toxic waste sites such as Love Canal and Times Beach in the 1970s. The Superfund Program allows the EPA to clean up such sites and to compel responsible parties to perform cleanups or reimburse the government for EPA-led cleanups.

2. The 2010 Census found that 7 percent identified themselves as black and the renter-occupied housing units made up 58 percent of all properties in the community.

3. Bourdieu clearly explicates the "downsizing" or "delocalization" effects of capitalism powered by increased financialization: "The increased freedom to

invest and, perhaps more crucially, to divest capital so as to obtain the highest financial profitability promotes the mobility of capital and the generalized delocalization of industrial or banking enterprises" (2003:93).

4. At the same time, like Bennett (2010), "I am ambivalent about Latour's claim that life (for Americans and Europeans) has simply become too technologized for the idea of pristine nature to wild any inspirational value . . . as the ideal of nature as the Wild continues to motivate some people to live more ecologically sustainable lives" (ibid.:121).

5. "IBM is committed to environmental affairs leadership in all of its business activities. IBM has had long-standing corporate policies of providing a safe and healthful workplace, protecting the environment, and conserving energy and natural resources." Retrieved at http://www.ibm.com/ibm/responsibility/policy9.shtml, on May 27, 2012.

6. Retrieved at http://www.ibm.com/ibm/responsibility/policy9.shtml, May 27, 2012. This business ethic is interesting because it deals strongly with "correcting" harm done, mistakes made. Ethical action, in this sense, is employed most firmly during post-disaster mitigation efforts when "responsibility" is linked to "prompt" action and state communication.

7. I had to re-distribute nearly a hundred of these surveys because I placed them in residents' mailboxes without proper postage. This did not please several carriers whose routes are in the plume.

8. Nobody I interviewed could tell me how IBM came up with this value, and many contested that a metric could be applied to the "loss" experienced by plume residents, which included losses that go beyond property values (e.g., ill health and a decaying quality of life).

9. The quantitative survey questionnaire was analyzed statistically, using the Statistical Package for the Social Sciences (SPSS). In addition to running frequencies and cross-tabulations to generate a general understanding of the survey results, I ran Kruskal-Wallis and Mann-Whitney tests (both nonparametric tests) to analyze the scaled questions to determine whether risk understandings differ between plume residents based on selected variables (e.g., gender, owner/renter status, age, etc.). In determining whether risk perception ratings are significantly different between plume residents, based on different control variables (e.g., renter vs. owner, age, gender, length of residence, etc.), the medians of the scaled (or ranked) responses were compared.

10. Neoliberalism is a term used to describe the political economic situation since the late 1970s, when the deregulation of financial markets became the norm, as did the decline in the bargaining power of labor unions as corporations, including IT firms like IBM, began offshoring and deindustrializing. Marking an aggressive departure from socialism at the end of the Cold War, neoliberalism began to unfold just about the time that both "globalization" entered public discourse and high-tech industry exploded on the global scene. IBM has always been an active ingredient of this so-called neoliberal computer revolution.

Along with other IT firms, IBM is an active player in providing information technology solutions to help monitor, track, and, ultimately, assist in the ongoing reformation of both globalized capitalism and the neoliberal state. They are coupled creations with shared interests in "devising cunning new ways of squeezing liquidity out of the fabulously complex flows of capital" (Besteman and Gusterson 2010:4).

NOTES TO CHAPTER 2

1. One could also look to the work of Ludwig Wittgenstein, who in his *Culture and Value* argued, "Work on philosophy—like work in architecture in many respects—is really more work on oneself. On one's own conception. On how one sees things. (And what one expects of them)" (Wittgenstein 1980 [1931], 24e).

2. As Stengers puts it, "Ecological practice (political in the broad sense) is then related to the production of values, to the proposal of new modes of evaluation, new meanings. But those values, modes of evaluation, and meanings do not transcend the situation in question, they do not constitute its intelligible truth. They are about the production of *new relations that are added* to a situation already produced by a multiplicity of relations" (Stengers 2010:33; emphasis in original).

3. Much of this scholarship was inspired by the foundational works of Henri Lefebvre (1991) and Gaston Bachelard (1958).

4. Legal theory, as Reno (2011) points out, matters when exploring the emplacement experience of residents coping with hazardous waste, especially since owners of properties in the plume possess "the legal ability—not just the physical might—to keep others [IBM] from interfering with one's acts" (Freyfogle 2010:80).

5. It is important here to draw attention to the observation that "Just as a privileging of ownership discourse obscures other commitments to and methods of caring for place, an emphasis on impaired enjoyment renders of environmental harm and justice or wider concerns about the consequences of neoliberal policies inadmissible" (Reno 2011:527).

6. See Downey and Dummit (1997), Gusterson (1996), Haraway (1989), Helmreich (1998), Martin (1994), Nader (1996), Rabinow (1999), Rajan (2006, 2012), Traweek (1988),

7. See Deleuze and Guattari (1987).

8. Retrieved at http://www.fema.gov/government/mitigation.shtm, June 7, 2012.

9. Habermas used the concept of "scientization" to describe this process, arguing that state actors turn "to strictly scientific recommendations in the exercise of their public functions" (Habermas 1970:62).

10. Risk, as earlier anthropological studies of risk have pointed out (Douglas and Wildavasky 1982; Douglas 1992, 1985), is a concept that has changed meaning over time and has been conditioned by a complex history of shifting social and institutional

conditions (Golding 1992). Originally employed to make sense of gambling in the seventeenth century, risk meant the probability of an event occurring in relation to the magnitude of losses or gains that might be entailed. In other words, since "risk" played a neutral role in the context of gambling, the very concept of "risk" was neutral (Hayes 1992). Today, however, risk is almost always associated with negativity or some negative outcome. Two milestone reports on risk—the Royal Society's 1983 report and the 1983 National Research Council (NRC 1983) report—highlight this conceptual emphasis on adversity. Risk, as it is defined in both reports, is "the probability that a particular adverse event occurs during a stated period of time, or results from a particular challenge" (Royal Society 1983:22).

11. Many of these studies implicitly or explicitly attend to social theoretical critiques of expertise, expert culture, and expert ideology put forth by philosophers like Jürgen Habermas and Michel Foucault.

12. "Partial connection" or "partial perspective" recognizes that in the act of knowledge production, the knower—scientist or otherwise—*disconnects* when practicing interpretation, yet never severs the tie between subjective and objective practice (Haraway 1991).

13. It is worth mentioning that suggesting our world is a "more modern" one might be problematic. For example, Latour contends that "we have never been modern" (Latour 1993). In that work, which is a good entry into the "anthropology of science," he proposed a rethinking of the Enlightenment idea of universal scientific truth, insisting that we think of facts not as ready-made, but facts that are fabricated and then represented as fact. This logic was used to question the very idea that humans actually live "modern" lives. Instead people, like facts, are under modernization, always in-the-making.

14. See Little 2012a for a slightly different approach to the emotion-contamination relationship, one that considers emotionality a front-and-center social dimension of disaster experience.

NOTES TO CHAPTER 3

1. History suggests technological developments in the United States "developed" from both private and public sector interests. For example, microchips, semiconductors, jet engines, personal computers, and the Internet all became high-growth industries through government research and development and support.

2. As German philosopher Peter Sloterdijk might put it, IBM is one of many firms of capitalism involved in the acceleration of cultural projects.

3. See Schumpeter (1942) for a detailed description of his use of the term "creative destruction" to explain both the effects and affects of capitalism driven by rapid innovation. According to Schumpeter, "Creative destruction is the essential fact about capitalism. It is what capitalism consists in and what every capitalist concern has got to live in ... Every piece of business strategy must be seen in its role in the perennial gale of creative destruction; it cannot be understood irrespective of it" (in McGraw 1997: 348; see also Zukin 1991). It is also critical

to remember that the logic of creative destruction is creatively recycled amidst contemporary corporate responsibility trends: "Corporate culture is an effort to create stability for the company that of course is never stable. The effort that [corporate responsibility] is, to change capitalism from a zero sum game, to deny or neutralize the destructive side of capitalism and to enhance its creativity or its ability for society and companies to work collaboratively, exacerbates the anxiety to perform culture, to make it measurable through business values, which in part fuels the kind of creative destruction of corporate activities that Schumpeter meant when he talked about capitalism" (McCarthy 2012:233–34).

4. In most of the parks G. F. Johnson developed in Endicott, Johnson City, and Binghamton, he built a carousel to be enjoyed by the public and free of charge. In fact, the "Triple Cities" were also nicknamed "Carousel City."

5. One of my sources, whose father worked for EJ, told me that "EJ was really concentrated on work shoes and this hurt them. It hurt because the shoe industry was moving toward stylish shoes that were more profitable and EJ wasn't doing that" (I.038).

6. Big Blue is a popular IBM nickname and there exist several theories of its origin. One theory is that IBM field representatives coined the term in the 1960s, referring to the color of the mainframes IBM installed in the 1960s and early 1970s. A second theory is that "All blue" was a term used to describe a loyal IBM customer, gaining popularity because of its use among business writers. A third theory suggests that Big Blue simply refers to IBM's company logo. A fourth theory suggests that Big Blue refers to a former company dress code that required many IBM employees to wear white shirts and blue suits.

7. IBM and National Geographic are collaborating on what is called the Genographic Project, an ambitious project using one of the largest collections of DNA samples ever assembled to map how the Earth was populated by humans (see Wade 2005).

8. In 2007, IBM announced the Spatiotemporal Epidemiological Modeler (STEM), a technology designed to enable the rapid creation of epidemiological models for how an infectious disease, such as avian influenza or dengue fever, is likely to geographically spread over time.

9. In FY 2008, Microsoft's revenues totaled $60.42 billion.

10. Apparently, Hollerith's original plan was to use strips of paper, but he soon found it was better to use a separate standardized card, which became known as a "punch card."

11. In May 2008, IBM announced its newest speedster nicknamed "Roadrunner," which was built for the Department of Energy's National Nuclear Security Administration and will be housed at Los Alamos National Laboratory in New Mexico. The IBM Roadrunner, which packs the power of 100,000 laptops, became the first computer to break a "petaflop" and will primarily be used, as stated on IBM's website, to "ensure national security, but will also help scientists perform research into energy, astronomy, genetics and climate change."

12. The so-called computerization of anthropology can be traced back to a symposium sponsored by the Wenner-Gren Foundation in 1962.

13. With THINK operating as its historical anchoring ethos, IBM could learn something important from the philosophers Giles Deleuze and Félix Guattari, who write: "To think is always to follow the witch's flight" (Deleuze and Guattari 1994:41). One could also turn to Hannah Arendt's *The Life of the Mind* where she explores an always relevant question regarding the doing of "thinking": "What are we 'doing' when we do nothing but think?" (Arendt 1981 [1971]: 8).

14. See Black (2001) for an in-depth look at IBM's lucrative business relationship with Nazi Germany.

15. IBM has been involved in many antitrust lawsuits over the years, with the most recent involving IBM's monopoly over the mainframe computer market of the computer industry. Richard DeLamarter (1986), in his book *Big Blue: IBM's Use and Abuse of Power*, draws on his work on antitrust suits while working for the U.S. Justice Department.

16. According to one source, state economic development aid helped to finance as much as 84 percent of the estimated $63 million purchase, which included IBM's Endicott real estate assets (see Yanchunas and Platsky 2002).

17. At the sale of IBM's microelectronics division, IBM agreed to keep 2,000 employees at the facility. As of 2011, there were roughly 600 IBMers working in this division. As many IBMers who got laid off during the 2002 closure will tell you, under EIT's mission to differentiate itself from IBM, it has slowly been getting rid of these leftover IBMers.

18. Retrieved June 28, 2012 from http://www.dec.ny.gov/regulations/2588.html.

19. One activist I interviewed considered this gift an example of local corruption and "collusion between IBM, local political officials, and the DEC." He jokingly added that all the money was "probably used to fix benches and tighten loose bolts in mayor's office," although he admitted not knowing exactly how it was spent. What we do know is that victims of toxic disasters commonly report a concern with state-corporate collusion in the aftermath of disaster (Button 2010).

20. The consent order is a dense official statement outlining the role and authority of the NYSDEC.

21. An IRM is a set of planned actions that can be conducted without extensive investigation and evaluation and is designed to be part of the final remedy for a contaminated site.

22. In situ thermal remediation is a process whereby electrical energy is introduced to the contaminated soil using a multitude of finite length cylindrical electrodes. The soil is heated up by the electrical current and the contaminated liquids and vapors are produced at the extraction well. By heating up the soil and vaporizing the contaminated liquid, it is anticipated that rate of extraction will increase and speed up VOC remediation efforts.

23. Retrieved at http://www.ibm.com/ibm/responsibility/communities.shtml, May 27, 2012.

NOTES TO CHAPTER 4

1. I employ the word "wane" here in remembrance of Claude Lévi-Strauss's book *A World on the Wane* (Lévi-Strauss 1961).

2. TCE is the most commonly used acronym for trichloroethylene, but trichloroethylene also goes by other names, including trichloroethene, ethylene trichloride, 1-chloro-2, 2-dichloroethylene, 1,1-dichloro-2-chloroethylene, 1,1,2-trichloroethylene, and TRI. TCE has also gone by the following trade names: Tri-clene, Dow-Tri, Germalgene, Westrosol, Flock-Flip, and Permachlor (see Harte et al. 1991).

3. The major players in this grassroots effort worked at the IBM plant in California's Silicon Valley, and their efforts have sparked a global sustainability and corporate accountability movement targeting the high-tech industry (Byster and Smith 2006).

4. Retrieved at www.tceblog.com. Accessed February 20, 2009.

5. Quantitative risk assessments, after all, have serious limitations because "Even with good data on actual human exposures. . . . the low signal-to-noise [or dose-response] ratios that typify epidemiological studies introduce considerable uncertainty when these data are introduced in practice" (Cullen and Small 2004:167).

6. Etymologically, "anecdote" comes from the Greek word *anekdota*, which translates as "unpublished memoirs" or "secret history." In practice, calling something "anecdotal" is to say it is unscientific or otherwise unreliable.

7. I think Barker's (2004) use of "oral testimony" is totally appropriate for anthropological studies of environmental health risk, for it seems to honor the fact that people do in fact witness things, and don't just perceive things. The latter has a chronic negative connotation, which has been the focus of the knowledge politics of the "risk society" (Beck 1992).

8. See Mandel et al. (2006) for a meta-analysis and review of the link between TCE and non-Hodgkins lymphoma.

9. The scaled survey responses were analyzed by running a Mann-Whitney test, which found statistically significant differences (p=.008) between renters and homeowners on this question.

10. See Hacker (2004, 2006).

11. Akers was president of IBM from 1983 to 1989, and CEO of IBM from 1984 until 1993. It should also be noted that Akers was on the Board of Directors of Lehman Brothers when it filed for bankruptcy in 2008.

12. The Housing Choice Voucher Program is a type of federal assistance provided by the U.S. Department of Housing and Urban Development dedicated to sponsoring subsidized housing for low-income families and individuals. It is more commonly known as Section 8, in reference to the portion of the U.S. Housing Act of 1937 under which the original subsidy program was authorized.

13. Activists in Victor, New York, the site of another TCE contaminated community, have successfully created a homeowners property protection plan. The extent to which this could work in Endicott and how people feel about the

efficacy of such a decision is unknown, and is likely a worthy topic of future ethnographic research.

14. This environmental research tends to draw on earlier theories of stigma developed by Goffman (1963), Ryan (1971), and Jones et al. (1984).

15. In my earlier fieldwork in Endicott, I visited two families in the plume that had reverse osmosis systems on their kitchen faucets (Little 2003).

NOTES TO CHAPTER 5

1. ITRC is funded primarily by the U.S. Department of Energy and Department of Defense, and the U.S. Environmental Protection Agency. It also receives funding from the ITRC Industry Affiliates Program (IAP).

2. Between 1986 and February 2004, the IBM-Endicott site was listed as a Class 4 site, meaning the site was determined to be properly closed, but that continued management was still required.

3. New York State assemblywoman, Donna Lupardo, helped pass this tenant notification bill (A10952B) in 2008 which aimed to "require property owners to provide notification of testing for contamination of indoor air to current and prospective tenants."

4. Activists in New York have pushed the state to reduce the TCE guideline from 10 micrograms per cubic meter to 5 micrograms per cubic meter, and the New York State Vapor Intrusion Alliance is working to lower that action level to meet levels set by states with more protective standards. Thus far, California, Colorado, and New Jersey have the most protective standards, which range from 0.016 to 0.02 mcg/m³.

5. The most common method of sampling indoor air contaminants where vapor intrusion is suspected or known is the use of a stainless steel sampling canister (e.g., Summa canister). Canister methods are most commonly used in North America (ITRC 2007).

6. He also told me that public access to these test results is a contentious issue that was raised with development of the tenant notification bill, but that data confidentiality issues had not been worked out yet.

7. One case study of a vapor intrusion mitigation program at the Redfield site in Denver, Colorado, has shown that VMSs do reduce risk (Folkes 2003; Folkes and Kurz 2002). Active depressurization systems—the same as those installed at the IBM-Endicott site—were been installed in more than 300 residential homes to mitigate indoor air concentrations of dichloroethene (DCE) resulting from migration of vapors from groundwater with elevated DCE concentrations. After monitoring the efficacy of these mitigation technologies over a three-year period, the data showed that these systems are capable of achieving the very substantial reductions in concentrations necessary to meet the concentration levels currently mandated by the state regulatory agency. Prior to installation of the system, 1,1-DCE indoor air concentrations ranged from below the reporting limit of 0.04 μg/m³ to over 100 μg/m³. Post-mitigation monitoring showed that in most cases, these systems were

able to reduce 1,1-DCE concentrations by 2 to 3 orders of magnitude, well below the state-required standards (Folkes 2003; Folkes and Kurz 2002). See USEPA (2008) for more details on the efficacy of these sub-slab depressurization systems.

8. The wide range in the order of magnitude here was an obvious source of frustration for activists I interviewed, primarily because they see an important difference between a risk ratio of 1 in a million and 1 in 10,000.

9. Bourdieu (1977:164) referred to *doxa* to mean the experience by which "the natural and social world appears as self-evident."

10. Edelstein (1988) draws on examples from several Superfund sites nationwide, including Love Canal (New York), Times Beach (Missouri), Legler (New Jersey), and Triana (Alabama) to demonstrate not only how the home becomes a prison once determined a toxic living space, but also a major source of financial loss, which in turn only exacerbates feelings of distress.

11. Many of my sources complained that Endicott residents pay too much in taxes. Because Endicott is a village within the town of Union, residents are required to pay village taxes, in addition to state and federal taxes.

12. IBM did offer homeowners in the plume $10,000, which they determined was about 8 percent of the property value lost as a result of the TCE contamination. All of my sources wonder how they came up with this figure.

13. As Siegel (2012:3) suggests, "Remediation, as well as mitigation, often helps property values recover. Routine mitigation, in which the installation of systems is not necessarily linked to findings of unacceptable exposures, may actually reverse stigma while protecting occupants from exposures to volatile organic compounds and radon." Residents in Endicott I spoke with never linked mitigation with the "recovery" of property values.

14. FY 2009 was marked by a $1.4 trillion deficit, or about 11.2 percent of the United States's GDP, according to the Congressional Budget Office. By February 2009, New York's deficit had reached $51 billion, the largest in the state's history. This has been described as "the fatal arithmetic of imperial decline" (Ferguson 2009:44).

NOTES TO CHAPTER 6

1. Political ecologist Jason Moore uses this "productivity and plunder" theme to describe the contentious interface of capitalism and environment. I learned of this phrasing during a lecture he gave at the Ferdinand Braudel Center at Binghamton University in 2009.

2. Harvey writes that "struggles against accumulation by dispossession are *fomenting* quite different lines of social and political struggle" (Harvey 2007:40; emphasis mine).

3. Allen (2003) draws on Alcoff (1991–1992) to formulate her "politics of enunciation."

4. Here, Bourdieu is actually referring to the social scientist, a social actor who he contends really "has no mandate, no mission, except those which he [*sic*] claims by virtue of the logic of his [*sic*] research" (1990:186). In this way, I think it's fair

to say that activists, like anthropologists, do what they do because they feel right doing it. In other words, they do it for themselves, even if they say they do it for others, for "the people."

5. My use of duty here is in line with Foucault's usage, which he discusses in his analysis of the Greek notion of *parrhesia*, or "frankness in speaking the truth": [In] *parrhesia*, telling the truth is regarded as a *duty*. The orator who speaks the truth to those who cannot accept his truth, for instance, and who may be exiled, or punished in some way, is *free* to keep silent. No one forces him to speak, but he feels that it is his duty to do so" (Foucault 2001 [1983]:19; emphasis in original).

6. Soon after I moved to Arizona in 2003, I stayed in contact with Allan Turnbull, who in 2004 developed the RAGE group. For whatever reason, Allan did not return my phone calls when I tried to interview him during my 2008–2009 field season. I was told by other activists and WBESC members that he had burned out or that he felt he had done his job and was therefore no longer active. Others told me that he dropped out of the spotlight for personal health reasons. Whatever the reasons for his current lack of involvement, his spirit and accomplishments are forever acknowledged and central to the storyline of toxics activism in Endicott.

7. One could also employ the term "relational subjectivity" to discuss intersubjectivity, as indicated in the work of McAfee (2000:129–50).

8. According to their website, Alliance@IBM/CWA Local 1701 is an IBM employee organization that is dedicated to preserving and improving the rights and benefits of IBM workers. "While our ultimate goal is collective bargaining rights with IBM, we will build our union now and challenge IBM on the many issues facing employees from off-shoring and job security to working conditions and company policy." Shelley contended that Alliance@IBM was "really a wannabe union. They want to be a union, but they're not."

9. Hillcrest, New York is another community struggling with TCE contamination in Broome County. I explain the experience of this community's most vocal stakeholder in a later section of this chapter.

10. IBM's founder, Thomas J. Watson, Sr., was born and raised in Potsdam, New York.

11. See the following New York State Department of Environmental Conservation link: http://www.dec.ny.gov/24.html.

12. I explain my own advocacy identity and engaged anthropological approach and intentions in chapter 8.

13. As depicted in Homer's *Odyssey IX*, lotus-eaters referred to a people who inhabited an island off the coast of north Africa that was lush with lotus plants, a plant known to cause sleepiness and intense apathy.

14. This comment reminded me of my own struggle as a researcher trying to blend in and navigate various social situations, to expose myself to the network of actors involved in the IBM pollution conflict and in vapor intrusion debates.

15. For an excellent in-depth journalistic account of the unfolding fracking issue in New York and Pennsylvania, see Wilber (2012).

16. Allen (2003) draws on Alcoff (1992) to formulate her "politics of enunciation."

17. Here, it should be noted, Bourdieu is actually referring to the social scientist, that social actor who "has no mandate, no mission, except those which he claims by virtue of the logic of his research" (1990, 186, [*sic*]).
18. Retrieved at http://www.louisvilleky.gov/apcd/star/. Accessed October 12, 2009.
19. Retrieved at www.ej4all.org. Accessed September 3, 2009.

NOTES TO CHAPTER 7

1. As Siegel (2009:2) has explained, "The most common attenuation factor is the ratio of the concentration of the gas in indoor air to the concentration of the gas in the subsurface source (soil gas, at some depth). It is usually labeled with the Greek letter *alpha* (α) and sometimes a subscript to show if it applies to an exterior soil-gas (sg) or a sub-slab (ss) sample."
2. The list of regulatory programs impacted by VI concerns also include Formerly Used Defense Sites (FUDS), Resource Conservation and Recovery Act sites (RCRA) and Hazardous and Solid Waste Amendment sites (HSWA), Underground Storage Tank sites (UST) and Above Ground Storage Tank sites (AST), and state-led cleanups.
3. Information on Kim Fortun's ongoing research project on "Tracking Exposure Science" can be found at http://kfortun.org/?page_id=43.
4. As Latour would have it, vapor intrusion is not "discovered," but instead "composed." Vapor intrusion science, in this sense, is what it currently is because of materialization, a vibrant mixture of historical ontology and political epistemology. In his words, "We need to have a much more material, much more mundane, much more immanent, much more realistic, much more embodied definition of the material world if we wish to compose a common world" (Latour 2010:484). Much of this re-materialization theory is inspired by the philosopher Alfred North Whitehead (see Whitehead 1978).
5. While Hess (1997a:162) is right when he contends that "I am better off positioning myself rather than letting someone else do it for me," this stance might be difficult to uphold for anthropologists and STS scholars trying to engage in emerging community-based participatory research or participatory action research projects.
6. Researchers at the Center for Wireless Integrated MicroSensing and Systems in the College of Engineering at the University of Michigan have more recently experimented with microsystem-based detection technologies called microfabricated gas chromatograph, which can detect TCE at very low levels, at the parts per trillion range. These sampling devices are the size of a wristwatch and are made of microfabricated silicon chips.

NOTES TO CHAPTER 8

1. According to Brown (2007:261), "Research silences are the questions we do not ask, the kinds of studies we do not do, the kinds of records we do not keep, the kind of implications we worry about making, the defense of beleaguered critical scientists that we fail to mount or join."

2. While not developed with deep care and attention here, Giorgio Agamben's notion of "bare life" (Agamben 1998) seems fitting in such sites of toxic struggle, as states and corporations tend to merge together in such disaster situations and in doing so simultaneously re-create and re-enforce the capacity of the sovereign power of the state and further marginalize concerned residents. The "bare life" of toxic struggle might be a useful framework to employ in emerging political ecologies and anthropologies of disaster and risk. Agamben might also be an important thinker to turn to for understanding what it means to act politically in an ambiguous mitigation zone.

3. As Žižek (2010) argues, amid global economic and ecological crises, there is a surplus in the presence of denial, anger, bargaining, depression, and acceptance. These observations link to the lived frustrations explored in this book. After all, amid all the traces of angst, are not Endicott residents, as well as the rest of us, equally exposed to or living in times of contentious repair and precarious mitigation? I explore this question in Little (2011).

4. Retrieved at http://www.ibm.com/ibm/responsibility/policy2.shtml, May 27, 2012.5. Retrieved at http://www.ibm.com/ibm/responsibility/environment.shtml, May 27, 2012.

6. As Rajak reminds us, "The discourse of CSR [corporate social responsibility] enables corporations to accrue moral authority as agents of progress and development, while simultaneously asserting their commitment to a global economic order governed by the supposedly amoral, asocial, and secular logic of 'the market'"(Rajak 2011:239).

7. Retrieved at http://www.ibm.com/smarterplanet/us/en/overview/ideas/index.html?re=sph, June 7, 2012.

8. Retrieved at http://www.ibm.com/smarterplanet/global/files/us__en_us__overview__decade_of_smart_011310.pdf, June 7, 2012.

9. Retrieved at http://www.ibm.com/ibm/responsibility/letter.shtml, May 27, 2012.

10. In his insightful *Adventures of Ideas*, Alfred North Whitehead (1933) reminds us that "In each period there is a general form of the forms of thought; and, like the air we breathe, such a form is so translucent, and so pervading, and so seemingly necessary, that only by extreme effort can we become aware of it" (Whitehead 1933:14). See Halewood (2011) for a deeper exploration of the critical influence of Whitehead's work in contemporary social theory.

11. See Whitehead (1926).

Adamson, J., M. M. Evans, and R. Stein, eds. 2002. *The Environmental Justice Reader: Politics, Poetics, Pedagogy*. Tucson: University of Arizona Press.

Agamben, Giorgio. 1998. *Homo Sacre: Sovereign Power and Bare Life*. D. Heller-Roazen, trans. Stanford: Stanford University Press.

Agency for Toxic Substances and Disease Registry. 2006. Historical Outdoor Air Emissions in Endicott, New York-International Business Machines Corporation (IBM). Summary Report. July.

Agyeman, Julian. 2005. *Sustainable Communities and the Challenge of Environmental Justice*. New York: New York University Press.

Agyeman, J., R. D. Bullard, and B. Evans, eds. 2003. *Just Sustainabilities: Development in an Unequal World*. Cambridge, MA: MIT Press.

Agyeman, J. and B. Evans. 2004. "Just Sustainability": The Emerging Discourse of Environmental Justice in Britain? *Geographical Journal* 170(2):155–64.

Alcoff, L. M. 1992. The Problem of Speaking for Others. *Cultural Critique* 20:5–32.

Allen, Barbara L. 2003. *Uneasy Alchemy: Citizens and Experts in Louisiana's Chemical Corridor Disputes*. Cambridge, MA: MIT Press.

Allen, Thomas J. and Michael S. Scott-Morton, eds.1994. *Information Technology and the Corporation of the 1990s*. New York: Oxford University Press.

Altman, Rebecca, et al. 2008. Pollution Comes Home and Pollution Gets Personal: Women's Experience of Household Toxic Exposure. *Journal of Health and Social Behavior* 49(4):417–35.

Arendt, Hannah. 1981 [1971]. *The Life of the Mind*. New York: Harcourt Brace.

Aswad, E. and S. M. Meredith. 2005. *IBM in Endicott*. Charleston, SC: Arcadia.

Auyero, Javier and Débora Alejandra Swistun. 2009. *Flammable: Environmental Suffering in an Argentine Shantytown*. New York: Oxford University Press.

———. 2008. Confused Because Exposed: Towards an Ethnography of Environmental Suffering. *Ethnography* 8(2):123–44.

———. 2007. The Social Production of Toxic Uncertainty. *American Sociological Review* 73:357–79.

Bachelard, Gaston. 1958. *The Poetics of Space*. Boston: Beacon Press.

Balshem, Martha. 1993. *Cancer in the Community: Class and Medical Authority*. Washington, DC: Smithsonian.

Barker, Holly M. 2004. *Bravo for the Marshallese: Regaining Control in a Post-Nuclear, Post-Colonial World*. Belmont, CA: Thomson and Wadsworth.

Basso, Keith and Steven Feld, eds. 1996. *Senses of Place*. Santa Fe, NM: School of American Research Press.

Batterbury, Simon and Leah Horowitz. Forthcoming. *Engaged Political Ecology*. Cambridge, MA: Open Book Publishers.

Beck, Ulrich. 2006. Living in the World Risk Society. *Economy and Society* 35(4):329–45.

———. 1998. *World Risk Society*. Cambridge: Polity Press.

———. 1994. The Reinvention of Politics: Towards a Theory of Reflexive Modernization. In U. Beck, A. Giddens, and S. Lash, eds., *Reflexive Modernization: Politics, Tradition and Aesthetics in the Modern Social Order*, pp. 1–55. Stanford: Stanford University Press.

———. 1992 [1986]. *Risk Society: Towards a New Modernity*. London: Sage.

Beck, Ulrich, Anthony Giddens, and Scott Lash. 1994. *Reflexive Modernization: Politics, Tradition and Aesthetics in the Modern Social Order*. Stanford: Stanford University Press.

Belden, Thomas G. and Marva R. Belden. 1962. *The Lengthening Shadow: The Life of Thomas J. Watson*. Boston: Little, Brown and Company.

Bennett, Jane. 2010. *Vibrant Matter: A Political Ecology of Things*. Durham and London: Duke University Press.

Berman, Marshall. 1982. *All That Is Solid Melts into Air: The Experience of Modernity*. New York: Penguin.

Bernard, Russell H. 2006. *Research Methods in Anthropology: Qualitative and Quantitative Approaches*. 4th ed. Lanham, MD: AltaMira.

Besteman, Catherine and Hugh Gusterson. 2010. Introduction. In H. Gusterson and C. Besteman, eds., *The Insecure American*, pp. 1–26. Berkeley: University of California Press.

Biersack, Aletta. 2006. Reimagining Political Ecology: Culture/Power/History/Nature. In A. Biersack and J. B. Greenberg, eds., *Reimaging Political Ecology*, pp. 3–42. Durham, NC: Duke University Press.

Black, Edwin. 2001. *IBM and the Holocaust*. New York: Three Rivers.

Blaikie, Piers and Harold Brookfield. 1987. *Land Degradation and Society*. London, UK: Methuen.

Bluestone, Barry and Bennett Harrison. 1982. *The Deindustrialization of America: Plant Closings, Community Abandonment, and the Dismantling of Basic Industry*. New York: Basic Books.

Borneman, John and Abdellah Hammoudi, eds. 2009. *Being There: The Fieldwork Encounter and the Making of Truth*. Berkeley: University of California Press.

Bourdieu, Pierre. 2003. *Firing Back: Against the Tyranny of the Market*. Loïc Wacquant, trans. New York: Verso.

———. 1991. *Language and Symbolic Power*. John B. Thompson, trans. Cambridge, MA: Harvard University Press.

———. 1990. *In Other Words: Essays towards a Reflexive Sociology.* Stanford: Stanford University Press.

———. 1977. *Outline of a Theory of Practice.* Richard Nice, trans. Cambridge, MA: Cambridge University Press.

———. 1963. *Travail et travailleurs en Algerie.* The Hague, Netherlands: Mouton.

Brock, Carla. 2009. Is Vapor Intrusion the Next Regulatory Juggernaut? *Seattle Daily Journal,* July 30:2.

Brown, Phil. 2007. *Toxic Exposures: Contested Illnesses and the Environmental Health Movement.* New York: Columbia University Press.

———. 2003. Qualitative Methods in Environmental Health Research. *Environmental Health Research* 111(14):1789–98.

———. 1992. Popular Epidemiology and Toxic Waste Contamination: Lay and Professional Ways of Knowing. *Journal of Health and Social Behavior* 33:267–81.

———. 1987. Popular Epidemiology: Community Response to Toxic Waste-Induced Illness in Woburn, Massachusetts. *Science, Technology, and Human Values* 12:78–85.

Brown, Phil., et al. 2004. Embodied Health Movements: Uncharted Territory in Social Movement Research. *Sociology of Health and Illness* 26:1–31.

Brown, Phil, Steve Kroll-Smith, and Valerie J. Gunter. 2000. Knowledge, Citizens, and Organizations. In Steve Kroll-Smith, Phil Brown, and Valerie J. Gunter, eds., *Illness and the Environment: A Reader in Contested Medicine,* pp. 9–25. New York: New York University Press.

Brown, Phil and Edwin Mikkelsen. 1997 [1990]. *No Safe Place: Toxic Waste, Leukemia, and Community Action.* Berkeley: University of California Press.

Bryant, B. 1995. *Environmental Justice: Issues, Policies, and Solutions.* Covelo: Island Press.

Bullard, Robert D. 2005. *The Quest for Environmental Justice: Human Rights and the Politics of Pollution.* San Francisco: Sierra Club Books.

———. 1994. *Dumping in Dixie: Race, Class, and Environmental Quality.* Boulder, CO: Westview Press.

———. 1993. *Confronting Environmental Racism: Voices from the Grassroots.* Boston: South End Press.

Burt, R. S. 1992. *Structural Holes: The Social Structure of Competition.* Cambridge, MA: Harvard University Press.

Burton, Michael L.1970. Computer Applications in Cultural Anthropology. *Computers and the Humanities* 5(1):37–45.

Büscher, Bram. 2010. Anti-Politics as Political Strategy: Neoliberalism and Transfrontier Conservation in Southern Africa. *Development and Change* 41(1):29–51.

Butler, Judith. 2011. Precarious Life: The Obligations of Proximity. The Neale Wheeler Watson Lecture. Nobel Museum, Svenska Akademiens Börssal. May 24.

Button, Gregory. 2010. *Disaster Culture: Knowledge and Uncertainty in the Wake of Human and Environmental Catastrophe.* Walnut Creek, CA: Left Coast Press.

Button, Gregory and Anthony Oliver-Smith. 2008. Disaster, Displacement, and Employment: Distortion of Labor Markets during Post-Katrina Reconstruction.

In Nandini Gunewardena and Mark Schuller, eds., *Capitalizing on Catastrophe: Neoliberal Strategies in Disaster Reconstruction*, pp. 123–46. Lanham, MD: AltaMira Press.

Byster, Leslie A. and Ted Smith. 2006. From Grassroots to Global: The Silicon Valley Toxics Coalition's Milestones in Building a Movement for Corporate Accountability and Sustainability in the High-Tech Industry. In T. Smith, D. A. Sonnenfeld, and D. N. Pellow, eds., *Challenging the Chip: Labor Rights and Environmental Justice in the Global Electronics Industry*, pp. 111–19. Philadelphia: Temple University Press.

Cable, Sherry, Tamara Mix, and Donald Hastings. 2005. Mission Impossible? Environmental Justice Activists' Collaborations with Professional Environmentalists and with Academics. In David Pellow and Robert Brulle, eds., *Power, Justice, and the Environment: A Critical Appraisal of the Environmental Justice Movement*, pp. 55–75. Cambridge, MA: MIT Press.

Cable, S. and T. Shriver. 1995. Production and Extrapolation of Meaning in the Environmental Justice Movement. *Sociological Spectrum* 15:419–42.

Callon, Michel. 1991. Techno-Economic Networks and Irreversibility. In John Law, ed., *A Society of Monsters: Essays on Power, Technology, and Domination*, pp. 132–62. New York: Routledge.

Capek, S. M. 1993. The "Environmental Justice" Frame: a Conceptual Discussion and Application. *Social Problems* 40(1): 5–24.

Caplan, Pat, ed. 2000. *Risk Revisited*. London: Pluto Press.

Carrier, James G. 2004. Introduction. In J. G. Carrier, ed., *Confronting Environments: Local Understanding in a Globalizing World*, pp. 1–29. Walnut Creek, CA: AltaMira.

Carson, Rachel. 1962. *Silent Spring*. Boston: Houghton Mifflin.

Casey, Edward. 1996. How to Get from Space to Place in a Fairly Short Stretch of Time: Phenomenological Prolegomena. In Keith Basso and Steven Feld, eds., *Sense of Place*, pp. 13–52. Santa Fe, NM: School of American Research Press.

Checker, Melissa. 2012. "Make Us Whole": Environmental Justice and the Politics of Skepticism. *Capitalism Nature Socialism* 23(3):35–51.

———. 2007. "But I Know It's True": Environmental Risk Assessment, Justice, and Anthropology. *Human Organization* 66(2):112–24.

———. 2005. *Polluted Promises: Environmental Racism and the Search for Justice in a Southern Town*. New York: New York University Press.

Chiu, Weihsueh A., et al. 2013. Human Health Effects of Trichloroethylene: Key Findings and Scientific Issues. *Environmental Health Perspectives* 121(3):303–11.

Clapp, Richard W. 2008. Cancer Mortality in IBM Endicott Plant Workers, 1969–2001: An Update on a NY Production Plant. *Environmental Health* 7(13):1–4.

———. 2006. Mortality among US Employees of a Large Computer Manufacturing Company: 1969–2001. *Environmental Health: A Global Access Science Source* 5:30.

———. 2002. Popular Epidemiology in Three Contaminated Communities. *Annals of the American Academy of Political and Social Science* 584:35–46.

Clarke, Simon. 2006. *From Enlightenment to Risk: Social Theory and Contemporary Society*. New York: Palgrave Macmillan.

Cliggett, Lisa, and Christopher A. Pool, eds. 2008. *Economies and the Transformation of Landscape*. Lanham, MD: AltaMira Press.

Cole, L. W. and S. R. Foster. 2001. *From the Ground Up: Environmental Racism and the Rise of the Environmental Justice Movement*. New York: New York University Press.

Corburn, Jason. 2005. *Street Science: Community Knowledge and Environmental Health Justice*. Cambridge, MA: MIT Press.

———. 2002. Combining Community-Based Research and Local Knowledge to Confront Asthma and Subsistence-Fishing Hazards in Greenpoint/Williamsburg, Brooklyn, New York. *Environmental Health Sciences* 110 (2):241–48.

Corvalán, C., D. Briggs, and G. Zielhuis, eds. 2000. *Decision-Making in Environmental Health: From Evidence to Action*. World Health Organization. London, UK: E. & F. N. Spon.

Couch, Stephan R. and Steve Kroll-Smith. 2000. Environmental Movements and Expert Knowledge: Evidence for a New Populism. In Steve Kroll-Smith, Phil Brown, Valerie J. Gunter, eds., *Illness and the Environment: A Reader in Contested Medicine*, pp. 384–404. New York: New York University Press.

Cowie, Jefferson. 1999. *Capital Moves: RCA's Seventy-Year Quest for Cheap Labor*. Ithaca, NY: Cornell University Press.

Cullen, Alison C. and Mitchell J. Small. 2004. Uncertain Risk: The Role and Limits of Quantitative Assessment. In T. McDaniels and M. J. Small, eds., *Risk Analysis and Society: An Interdisciplinary Characterization of the Field*, pp. 163–212. Cambridge, MA: Cambridge University Press.

Cummings, R. B. 1981. Is Risk Assessment a Science? *Risk Analysis* 1(1):1–3.

Cutter, S. 1995. Race, Class and Environmental Justice. *Progress in Geography* 19(1):111–22.

Cutter, S. L., J. Tiefenbacher, and W. D. Solecki. 1992. En-gendered Fears: Femininity and Technological Risk Perception. *Industrial Crisis Quarterly* 6:5–22.

Davidson, D. and W. Freudenburg. 1996. Gender and Environmental Risk Concerns: A Review and Analysis of Available Research. *Environment and Behavior* 28:302–39.

de Certeau, Michel. 1984. *The Practice of Everyday Life*. Steven Rendall, trans. Berkeley: University of California Press.

DeLamarter, Richard. 1986. *Big Blue: IBM's Use and Abuse of Power*. New York: Dodd Mead.

Deleuze, Gilles. 1992. Postscript on the Societies of Control. *October* 59:3–7.

Deleuze, Gilles and Félix Guattari. 1994 [1991]. *What is Philosophy?* Hugh Tomlinson and Graham Burchell, trans. New York: Columbia University Press.

———. 1987. *A Thousand Plateaus: Capitalism and Schizophrenia*. Brian Massumi, trans. Minneapolis: University of Minnesota Press.

De Waal, Alexander. 2008. Foreword. In Nandini Gunewardena and Mark Schuller, eds., *Capitalizing on Catastrophe: Neoliberal Strategies in Disaster Reconstruction*, pp. ix–xiv. Lanham, MD: Alta Mira.

Dewey, John. 2005 [1910]. *How We Think*. Barnes and Noble: New York.

Di Chiro, G. 1998. Environmental Justice from the Grassroots: Reflections on History, Gender, and Expertise. In D. Faber, ed., *The Struggle for Ecological Democracy:*

Environmental Justice Movements in the United States, pp. 104–36. New York: Guilford Press.

———. 1995. Nature as Community: The Convergence of Environment and Social Justice. In W. Cronon, ed., *Uncommon Ground: Toward Reinventing Nature*, pp. 298–320. New York: Norton.

Douglas, Mary. 1993. The Idea of a Home: A Kind of Space. In A. Mack, ed., *Home: A Place in the World*, pp. 261–81. New York: New York University Press.

———. 1992. *Risk and Blame: Essays in Cultural Theory*. New York: Routledge.

———. 1985. *Risk Acceptability According to the Social Sciences*. New York: Russell Sage.

Douglas, Mary and Aaron Wildavsky. 1982. *Risk and Culture: An Essay on the Selection of Technical and Environmental Dangers*. Berkeley: University of California Press.

Downey, Gary Lee and Joseph Dumit. 1997. *Cyborgs and Citadels: Anthropological Interventions in Emerging Sciences and Technologies*. Santa Fe, NM: School of American Research Press.

Dowty, Rachel A. and Barbara L. Allen. 2011. *Dynamics of Disaster: Lessons on Risk, Response, and Recovery*. London: Earthscan.

Dunn, C. and S. Kingham. 1996. Establishing Links between Air Quality and Health: Searching for the Impossible? *Social Science Medicine* 42 (6):831–41.

Dyer-Witherford, Nick. 1999. *Cyber-Marx: Cycles and Circuits in High-Tech Capitalism*. Urbana: University of Illinois Press.

Edelman, Marc. 2001. Social Movements: Changing Paradigms and Forms of Politics. *Annual Review of Anthropology* 30:285–317.

Edelstein, M. R. and W. J. Makofske. 1998. *Radon's Deadly Daughters: Science, Environmental Policy, and the Politics of Risk*. New York: Rowman and Littlefield.

Edelstein, Michael R. 2004. *Contaminated Communities: Coping with Residential Toxic Exposure*. Boulder, CO: Westview.

———. 2000. "Outsiders Just Don't Understand": Personalization of Risk and the Boundary Between Modernity and Postmodernity. In M. Cohen, ed., *Risk in the Modern Age: Social Theory, Science, and Environmental Decision-Making*, pp.123–42. New York: St. Martin's Press.

———. 1993. When the Honeymoon Is Over: Environmental Stigma and Distrust in the Siting of a Hazardous Waste Disposal Facility in Niagara Falls, New York. *Research in Social Problems and Public Policy* 5:75–96.

———. 1992. NIMBY as a Healthy Response to Environmental Stigma Associated with Hazardous Facility Siting. In G. Leitch, ed., *Hazardous Material/Wastes: Social Aspects of Facility Planning and Management*, pp. 413–31. Winnipeg, Manitoba: Institute for Social Impact Assessment.

———. 1991. Ecological Threats and Spoiled Identities: Radon Gas and Environmental Stigma. In S. Couch and S. Kroll-Smith, eds., *Communities at Risk: Community Responses to Technological Hazards*, pp. 205–26. New York: Peter Lang.

———. 1988. *Contaminated Communities: The Social and Psychological Impacts of Residential Toxic Exposure*. Boulder, CO: Westview.

———. 1987. Toward a Theory of Environmental Stigma. In J. Harvey and D. Henning, eds., *Public Environments*, pp. 21–25. Ottawa, Canada: Environmental Design Research Association.

———. 1986. Toxic Exposure and the Inversion of Home. *Journal of Architecture and Planning Research* 3:237–51.

———. 1984. *Stigmatizing Aspects of Toxic Pollution*. Report prepared for the law firm Martin & Snyder for *Cito v. Monsanto*.

———. 1981. *The Social and Psychological Impacts of Groundwater Contamination in the Legler Section of Jackson, New Jersey*. Report prepared for law firm Kreindler & Kreindler for *Ayers v. Jackson Township*.

Emerson, Robert et al. 1995. *Writing Ethnographic Fieldnotes*. Chicago: Chicago University Press.

Englund, Harri. 2002. Ethnography after Globalism: Migration and Emplacement in Malawi. *American Ethnologist* 29 (2):261–86.

Epstein, B. 1997. The Environmental Justice/Toxics Movement: Politics of Race and Gender. *Capitalism, Nature, Socialism* 8 (3):63–87.

Erickson, B. E. 2007. Toxin or Medicine? Explanatory Models of Radon in Montana Health Mines. *Medical Anthropology Quarterly* 21(1):1–21.

Erikson, Kai. 1994. *A New Species of Trouble: Explorations in Disaster, Trauma, and Community*. New York: Norton.

———. 1991. A New Species of Trouble. In Stephan Crouch and J. Stephen Kroll-Smith, eds., *Communities at Risk: Community Responses to Technological Disasters*, pp. 12–29. New York: Peter Lang.

Escobar, Arturo. 1999. After Nature: Steps to an Antiessentialist Political Ecology. *Current Anthropology* 40:1–30.

———. 1998. Whose Knowledge, Whose Nature? Biodiversity, Conservation, and the Political Ecology of Social Movements. *Journal of Political Ecology* 5:53–82.

———. 1996. Constructing Nature: Elements for a Poststructural Political Ecology. In Richard Peet and Michael Watts, eds., *Liberation Ecologies: Environment, Development, Social Movements*. pp. 46–68. London and New York: Routledge.

Faber, Daniel, ed. 1998. *The Struggle for Ecological Democracy: Environmental Justice Movements in the United States*. New York: Guilford Press.

Fairhead, James and Ian Scoones. 2005. Local Knowledge and the Social Shaping of Soil Investments: Critical Perspectives on the Assessment of Soil Degradation in Africa. *Land Use Policy* 22:33–41.

Farmer, Paul. 1999. *Infections and Inequalities: The Modern Plagues*. Berkeley: University of California Press.

Ferguson, James. 1994. *The Anti-Politics Machine. "Development," Depoliticization, and Bureaucratic Power in Lesotho*. Minneapolis: University of Minnesota Press.

Ferguson, Niall. 2009. Empire at Risk. *Newsweek*, December 7.

Fischer, Frank. 2003. *Citizens, Experts, and the Environment: The Politics of Local Knowledge*. Durham, NC: Duke University Press.

Fischer, Michael D. 1994. *Applications in Computing for Social Anthropologists*. New York: Routledge.

Fischhoff, Baruch. 1995. Risk Perception and Communication Unplugged: Twenty Years of Process. *Risk Analysis* 15 (2):137–45.

Fitzpatrick, Kevin and Mark LaGory. 2000. *Unhealthy Places: The Ecology of Risk in the Urban Landscape*. New York: Routledge.

Flynn, James, Paul Slovic, and C. K. Mertz. 1994. Gender, Race, and Perception of Environmental Health Risks. *Risk Analysis* 14 (6):1101–8.

Folkes, David. 2003. *Design, Effectiveness, and Reliability of Sub-slab Depressurization Systems*. EPA Seminar on Indoor Air Vapor Intrusion. Atlanta, February 25–26. Retrieved at http://www.envirogroup.com/publications/folkes_epa_seminar.pdf. Accessed August 13, 2009.

Folkes, David et al. 2009. Observed Spatial and Temporal Distributions of CVOCs at Colorado and New York Vapor Intrusion Sites. *Ground Water Monitoring and Remediation* 29 (1):70–80.

Folkes, D. J. and D. W. Kurz. 2002. *Efficacy of Subslab Depressurization for Mitigation of Vapor Intrusion of Chlorinated Organic Compounds*. Proceedings of Indoor Air 2002. Retrieved at http://www.envirogroup.com/publications/effi cancyofslab.pdf. Accessed August 28, 2009.

Foran, Tira and David A. Sonnenfeld. 2006. Corporate Social Responsibility in Thailand's Electronics Industry. In T. Smith, D. A. Sonnenfeld, D. N. Pellow, eds., *Challenging the Chip: Labor Rights and Environmental Justice in the Global Electronics Industry*. pp. 70–82. Philadelphia: Temple University Press.

Forand, Steven P., Elizabeth L. Lewis-Michl, and Marta I. Gomez. 2012. Adverse Birth Outcomes and Maternal Exposure to Trichloroethylene and Tetrachloroethylene through Soil Vapor Intrusion in New York State. *Environmental Health Perspectives* 120 (4):616–21.

Fortun, Kim. 2012. Ethnography in Late Industrialism. *Cultural Anthropology* 27(3):446-64.

2001. *Advocacy after Bhopal: Environmentalism, Disaster, New Global Orders*. Chicago: University of Chicago Press.

Fortun, Mike and Herbert J. Bernstein.1998. *Muddling Through: Pursuing Science and Truths in the 21st Century*. Washington, DC: Counterpoint.

Fortun, Mike and Kim Fortun. 2005. Scientific Imaginaries and Ethical Plateaus in Contemporary US Toxicology. *American Anthropologist* 107(1):43–54.

Foster, John Bellamy and Fred Magdoff. 2009. *The Great Financial Crisis: Causes and Consequences*. New York: Monthly Review.

Foucault, Michel. 2001 [1983]. *Fearless Speech*. Joseph Pearson, ed. Los Angeles: Semiotext(e).

Fox, S.1991. *Toxic Work: Women Workers at GTE Lenkurt*. Philadelphia: Temple University Press.

Foy, Nancy. 1975. *The Sun Never Sets on IBM*. New York: William Morrow.

Franklin, Sarah. 1995. Science as Culture, Cultures of Science. *Annual Review of Anthropology* 24:163–84.

Freudenberg, Nicholas. 1984. *Not in Our Backyards! Community Action for Health and the Environment.* New York: Monthly Review Press.

Freudenburg, W. R. and T. Jones. 1991. Attitudes and Stress in the Presence of Technological Risk: A Test of the Supreme Court Hypothesis. *Social Forces* 69(4):1143–68.

Freyfogle, Eric T. 2010. Property and Liberty. *Harvard Environmental Law Review* 34:75–118.

Frickel, Scott and M. Bess Vincent. 2011. Katrina's Contamination: Regulatory Knowledge Gaps in the Making and Unmaking of Environmental Contention. In Rachel A. Dowty and Barbara L. Allen, eds., *Dynamics of Disaster: Lessons on Risk, Response and Recovery*, pp. 11–28. London: Earthscan.

Geertz, Clifford. 2000. *Available Light: Anthropological Reflections on Philosophical Topics.* Princeton, NJ: Princeton University Press.

———. 1983. *Local Knowledge: Further Essays in Interpretive Anthropology.* New York: Basic Books.

Giddens, Anthony. 1990. *The Consequences of Modernity.* Cambridge, MA: Polity Press.

———. 1991. *Modernity and Self-Identity: Self and Society in the Late Modern Age.* Cambridge, MA: Polity Press.

Goffman, Erving. 1963. *Stigma: Notes on the Management of Spoiled Identities.* Englewood Cliffs, NJ: Prentice-Hall.

Golding, Dominic. 1992. A Social and Programmatic History of Risk Research. In Sheldon Krimsky and Dominic Golding, eds., *Social Theories of Risk*m pp. 23–52. London: Praeger.

Goldman, Mara J., Paul Nadasdy, and Matthew D. Turner, eds. 2011. *Knowing Nature: Conversations at the Intersection of Political Ecology and Science Studies.* Chicago and London: University of Chicago Press.

Gottlieb, Robert. 2005 [1993]. *Forcing the Spring: The Transformation of the American Environmental Movement.* Washington, DC: Island Press.

Graeber, David. 2010. Neoliberalism, or the Bureaucratization of the World. In H. Gusterson and C. Besteman, eds., *The Insecure American*, pp. 79–96. Berkeley: University of California Press.

———. 2009. *Direct Action: An Ethnography.* Edinburgh, UK: AK Press.

Gramza, Janet. 2009. Life in the Plume: IBM's Pollution Haunts a Village. *The Post-Standard.* January 11.

Greenwood, Jeremy. 1997. *The Third Industrial Revolution.* American Enterprise Institute, August. Washington, DC: AEI Press.

Grossman, Elizabeth. 2006. *High Tech Trash: Digital Devices, Hidden Toxics, and Human Health.* Washington, DC: Island Press.

Gustafson, Per E. 1998. Gender Differences in Risk Perception: Theoretical and Methodological Perspectives. *Risk Analysis* 18(6):805–11.

Gusterson, Hugh. 1996. *Nuclear Rites: A Weapons Laboratory at the End of the Cold War*. Berkeley: University of California Press.

Habermas, J. 1970. *Toward a Rational Society: Student Protest, Science, and Politics*. Jeremy Shapiro, trans. Boston: Beacon Press.

Hacker, Jacob. 2006. *The Great Risk Shift*. New York and Oxford: Oxford University Press.

———. 2004. *The Privatization of Risk and the Growing Economic Insecurity in America*. Social Science Research Council forum. Retrieved at http://privatizationofrisk.ssrc.org/Hacker/. Accessed May 15, 2009.

Hacking, Ian. 1999. *The Social Construction of What?* Cambridge, MA: Harvard University Press.

Haenn, Nora and David G. Casagrande. 2007. *Citizens, Experts, and Anthropologists: Finding Paths in Environmental Policy*. Human Organization 66(2):99–102.

Halewood, Michael. 2011. *A. N. Whitehead and Social Theory: Tracing a Culture of Thought*. London: Anthem Press.

Hammersley, Martyn and Paul Atkinson. 1995. *Ethnography: Principles in Practice*. London: Routledge.

Haraway, Donna. 2011. Love in a Time of Extinctions and Exterminations: Staying with the Trouble. Wellek Library Lecture. Critical Theory Institute. University of California, Irvine. May 3.

———. 1997. *Modest_Witness@Second_Millenium*. New York: Routledge.

———. 1991. *Simians, Cyborgs, and Women: The Reinvention of Nature*. New York: Routledge.

———. 1989. *Primate Visions: Gender, Race, and Nature in the World of Modern Science*. New York: Routledge.

———. 1988. Situated Knowledges: The Science Question in Feminism and the Privilege of Partial Perspective. *Feminist Studies* 14(3):575–99.

Harding, Sandra. 1991a. Who Knows? Identities and Feminist Epistemology. In Joan E. Hartman and Ellen Messer-Davidow, eds., *(En)Gendering Knowledge: Feminists in Academe*. pp. 100–15. Knoxville: University of Tennessee Press.

———. 1991b. *Whose Science, Whose Knowledge?* Ithaca, NY: Cornell University Press.

Harper, Janice. 2002. *Endangered Species: Health, Illness and Death among Madagascar's People of the Forest*. Durham, NC: Carolina Academic Press.

Harte, John et al. 1991. *Toxics A to Z: A Guide to Everyday Pollution Hazards*. Berkeley: University of California Press.

Harvey, David. 2007. Neoliberalism as Creative Destruction. *Annals of the American Academy of Political and Social Science* 610(21):22–44.

———. 2005. *A Brief History of Neoliberalism*. New York: Oxford University Press.

———. 1996. *Justice, Nature, and the Geography of Difference*. Oxford, UK: Blackwell.

———. 1985. The Geopolitics of Capitalism. In D. Gregory and J. Urry, eds., *Social Relations and Spatial Structures*. pp.128–63. London: Macmillan.

Hastrup, Kirsten. 1995. *A Passage to Anthropology: Between Experience and Theory*. New Haven, CT: Yale University Press.

Hawes, Amanda and David N. Pellow. 2006. The Struggle for Occupational Health in Silicon Valley: A Conversation with Amanda Hawes. In T. Smith, D. A. Sonnenfeld, D. N. Pellow, eds., *Challenging the Chip: Labor Rights and Environmental Justice in the Global Electronics Industry*, pp. 120–28. Philadelphia: Temple University Press.

Hayes, Michael V. 1992. On the Epistemology of Risk: Language, Logic, and Social Science. *Social Science and Medicine* 35(4):401–7.

Heidegger, Martin. 1968. *What is Called Thinking?* J. Glenn Gray, trans. New York: Harper & Row.

Helmreich, Stefan. 1998. *Silicon Second Nature: Culturing Artificial Life in a Digital World*. Berkeley: University of California Press.

Hess, David J. 2007a. *Alternative Pathways in Science And Industry: Activism, Innovation, and the Environment in an Era of Globalization*. Cambridge, MA: MIT Press.

———. 2007b. Crosscurrents: Social Movements and the Anthropology of Science and Technology. *American Anthropologist* 109(3):463–72.

———. 2006. Antiangiogenesis Research and the Dynamics of Scientific Fields. In S. Frickel and K. Moore, eds., *The New Political Sociology of Science: Institutions, Networks and Power*. Madison: University of Wisconsin Press.

———. 1997a. If You're Thinking of Living in STS: A Guide for the Perplexed. In G. L. Downey and J. Dumit, eds., *Cyborgs and Citadels*, pp. 143–64. Santa Fe, NM: School of American Research Press.

———. 1997b. *Science Studies: An Advanced Introduction*. New York: New York University Press.

———. 1992. Introduction: The New Ethnography and the Anthropology of Science and Technology. In David Hess and Linda Layne, eds., *Knowledge and Society 9: The Anthropology of Science and Technology*, pp. 1–26. Greenwich, CT: JAI Press.

Hess, David and Linda Layne, eds. 1992. *Knowledge and Society 9: The Anthropology of Science and Technology*. Greenwich, CT: JAI Press.

Hinchliffe, Stephen. 2007. *Geographies of Nature: Societies, Environments, Ecologies*. London: Sage.

Hines, Christine. 2000. *Virtual Ethnography*. London: Sage.

Hiskes, Richard P. 1998. *Democracy, Risk, and Community: Technological Hazards and the Evolution of Liberalism*. New York: Oxford University Press.

Hofrichter, Richard, ed. 2000. *Reclaiming the Environmental Debate: The Politics of Health in a Toxic Culture*. Cambridge, MA: MIT Press.

———. 1993. *Toxic Struggles: The Theory and Practice of Environmental Justice*. Philadelphia: New Society.

Holifield, Ryan. 2004. Neoliberalism and Environmental Justice in the United States Environmental Protection Agency: Translating Policy into Managerial Practice in Hazardous Waste Remediation. *Geoforum* 35:285–97.

Hopkins, M. 2003. *The Planetary Bargain: Corporate Social Responsibility Matters*. London: Earthscan.

Horowitz, Leah S. 2008. "It's Up to the Clan to Protect": Cultural Heritage and the Micropolitical Ecology of Conservation in New Caledonia. *Social Science Journal* 45:258–78.

Hymes, Dell, ed. 1965. *The Use of Computers in Anthropology*. The Hague, Netherlands: Mouton.

Illich, Ivan. 1977. Disabling Professions. In I. Illich et al., eds., *Disabling Professions*. London: Marion Boyers.

Inglis, William. 1935. *George Johnson and His Industrial Democracy*. New York: Huntington.

Inglod, Tim. 2000. *The Perception of the Environment: Essays in Livelihood, Dwelling and Skill*. New York: Routledge.

International Business Machines Corporation. 2008a. IBM Corporate Archive. Document code 9215FQ14. Retrieved at http://www.ibm.com. Accessed October 20, 2008.

International Business Machines Corporation. 2008b. IBM Corporate Archive. Document code 2410MP03. Retrieved at http://www.ibm.com. Accessed October 20, 2008.

International Business Machines Corporation. 2008c. IBM Corporate Archive. Document code 9215FQ14. Retrieved at http://www.ibm.com. Accessed October 20, 2008.

Interstate Technology and Regulatory Council. 2007. *Vapor Intrusion Pathway: A Practical Guideline*. ITRC Technical and Regulatory Guidance Document. January.

Irwin, Alan. 1997. *Citizen Science: A Story of People, Expertise and Sustainable Development*. New York: Routledge.

Irwin, Alan, Alison Dale, and Denis Smith. 1996. Science and Hell's Kitchen: The Local Understanding of Hazard Issues. In Alan Irwin and Brian Wynne, eds., *Misunderstanding Science? The Public Reconstruction of Science and Technology*, pp. 47–64. New York: Cambridge University Press.

Irwin, Alan and Brian Wynne. 1996. *Misunderstanding Science? The Public Reconstruction of Science and Technology*. New York: Cambridge University Press.

Jackson, Deborah D. 2011. Scents of Place: The Dysplacement of a First Nations Community in Canada. *American Anthropologist* 113(4):606–18.

Jas, Nathalie, and Soraya Boudia, eds. 2011. *Powerless Science? The Making of the Toxic World in the Twentieth Century*. New York: Berghahn Books.

Jasanoff, Sheila. 2004. The Idiom of Co-Production. In Sheila Jasanoff, ed., *State of Knowledge: The Co-Production of Science and Social Order*, pp. 1–12. New York: Routledge.

———. 1998. The Political Science of Risk Perception. *Reliability Engineering and Systems Safety* 59:91–99.

———. 1992. Science, Politics, and the Renegotiation of Expertise at EPA. *Osiris* 7:195–217.

———. 1990. *The Fifth Branch: Science Advisors as Policymakers*. Cambridge, MA: Harvard University Press.

Jewitt, Sarah. 2008. Political Ecology of Jharkhand Conflicts. *Asia Pacific Viewpoint* 49(1):68–82.

Jones, Edward, et al. 1984. *Social Stigma: The Psychology of Marked Relationships*. New York: Freeman.

Jonscher, Charles. 1994. An Economic Study of the Information Technology Revolution. In T. J. Allen and M.S. Scott-Morton, eds., *Information Technology and the Corporation of the 1990s*, pp. 5–42. New York: Oxford University Press.

Kleinman, Arthur. 1980. *Patients and Healers in the Context of Culture: An Exploration of the Borderland between Anthropology, Medicine, and Psychiatry*. Berkeley: University of California Press.

Knorr-Cetina, K. 1999. *Epistemic Cultures: How the Sciences Make Knowledge*. Cambridge, MA: Harvard University Press.

Kroll-Smith, J. S. and S. Couch. 1993. Symbols, Ecology, and Contamination: Case Studies in the Ecological-Symbolic Approach to Disaster. *Research in Social Problems and Public Policy* 5:47–73.

———. 1990. *The Real Disaster is Above Ground: A Mine Fire and Social Conflict*. Lexington: University of Kentucky Press.

Kuletz, Valerie L. 1998. *The Tainted Desert: Environmental and Social Ruin in the American West*. New York: Routledge.

Kurtz, H. 2002. The Politics of Environmental Justice as the Politics of Scale: St. James Parish, Louisiana, and the Shintech Siting Controversy. In A. Herod and M. W. Wright, eds., *Geographies of Power: Placing Scale*, pp. 249–73. Oxford, UK: Basil Blackwell.

LaDou, Joseph. 2006. Occupational Health in the Semiconductor Industry. In Ted Smith, David A. Sonnenfeld, and David N. Pellow, eds., *Challenging the Chip: Labor Rights and Environmental Justice in the Global Electronics Industry*, pp. 31–42. Philadelphia: Temple University Press.

———. 1994. Health Issues in the Global Semiconductor Industry. *Annals of the Academy of Medicine* [Singapore] 23, no. 5 (September).

———. 1984. The Not-So-Clean Business of Making Chips. *Technology Review* 87, no. 4 (May–June).

———. 1983. Potential Occupation Health Hazards in the Microelectronics Industry. *Scandinavian Journal of Work, Environment and Health* 9(1):42–46.

LaDou, Joseph and T. Rohm. 1998. The International Electronics Industry. *International Journal of Occupational Medicine* 4:1–18.

Lake, R. W. 1996. Volunteers, Nimbys, and Environmental Justice: Dilemmas of Democratic Practice. *Antipode* 28(2):160–74.

Latour, Bruno. 2010a. Networks, Societies, Spheres: Reflections of an Actor-Network Theorist. Keynote speech, *International Seminar on Network Theory: Network Multidimensionality in the Digital Age*. February 9. Annenberg School of Communication and Journalism. Los Angeles.

———. 2010b. An Attempt at a "Compositionist Manifesto." *New Literary History* 41:471–90.

——. 2007. How to Think Like a State. WRR Lecture 2007-*The Thinking State?* Scientific Council for Government Policy. November 22. The Hague, Netherlands.

——. 2005. *Reassembling the Social: An Introduction to Actor-Network-Theory.* New York: Oxford University Press.

——. 2004. *Politics of Nature: How to Bring the Sciences into Democracy.* Cambridge, MA: Harvard University Press.

——. 1998. To Modernize or Ecologize? That is the Question. In Bruce Braun and Noel Castree, eds., *Remaking Reality: Nature at the Millennium,* pp. 221–42. New York: Routledge.

——. 1993. *We Have Never Been Modern.* Catherine Porter, trans. Cambridge, MA: Harvard University Press.

——. 1987. *Science in Action: How to Follow Scientists and Engineers through Society.* Cambridge, MA: Harvard University Press.

Latour, Bruno. 1987. *Science in Action.* Milton Keynes: Open University Press.

Latour, Bruno and Steve Woolgar. 1979. *Laboratory Life: The Social Construction of Scientific Facts.* Beverly Hills, CA: Sage.

Lefebvre, Henri. 1991 [1974]. *The Production of Space.* Donald Nicholson-Smith, trans. Oxford, UK: Basil Blackwell.

Lerner, Stephen D. 2010. *Sacrifice Zones: The Front Lines of Toxic Chemical Exposure in the United States.* Cambridge, MA: MIT Press.

Levine, Adeline. 1982. *Love Canal: Science, Politics, and People.* Boston: Lexington.

Lévi-Strauss, Claude. 1961. *A World on the Wane.* John Russell, trans. New York: Criterion.

Lind, N. C. 1987. Is Risk Analysis an Emerging Profession? *Risk Abstracts* (4)4:167–69.

Little, Peter C. 2013a. Envisioning the Political Ecology of Mitigation in a Microelectronic Disaster Setting. *Journal of Political Ecology* 20(13):217–37.

——. 2013b. Vapor Intrusion: The Political Ecology of an Emerging Environmental Health Concern. *Human Organization* 72(2):121–31.

——. 2012a. Another Angle on Pollution Experience: Toward an Anthropology of the Emotional Ecology of Risk Mitigation. *Ethos* 40(4):431–52.

——. 2012b. Environmental Justice Discomfort and Disconnect in IBM's Tainted Birthplace: A Micropolitical Ecology Perspective. *Capitalism Nature Socialism* 23(2):92–109.

——. 2011. Living in Mitigated Times in IBM's Tainted Birthplace. Paper presented at the American Anthropological Association Annual Meeting. Montreal, Quebec. November 17.

——. 2010. "Instead of Tumbleweed, We Have Mitigation Systems": An Ethnography of Toxics Risk, Mitigation, and Advocacy in IBM's Deindustrialized Birthplace. PhD diss., Department of Anthropology, Oregon State University.

——. 2009. Negotiating Community Engagement and Science in the Federal Environmental Public Health Sector. *Medical Anthropology Quarterly* 23(2):94–118.

——. 2003. *Political Ecology, Health, and Environment: Groundwater Contamination in Upstate New York and Its Surfacing Narratives.* Unpublished honors thesis, Department of Anthropology, Binghamton University.

Lockheed Martin Corporation. Systems Integration-Owego. Retrieved at http://www. lockheedmartin.com/si/. Accessed November 3, 2008.

Lohr, Steve. 2008. Why Old Technologies Are Still Kicking. *New York Times*, March 23.

Low, Setha M. and Denise Lawrence-Zúñiga. 2003. *The Anthropology of Space and Place: Locating Culture*. Oxford, UK: Blackwell.

Lupton, Deborah. 1999. *Risk*. New York: Routledge.

Mandel, J. H. et al. 2006. Occupational Trichloroethylene Exposure and Non-Hodgkin's Lymphoma: A Meta-Analysis and Review. *Occupational and Environmental Medicine* 63:597–607.

Maney, Kevin, Steve Hamm, and Jefferey M. O'Brien. 2011. *Making the World Work Better: The Ideas that Shaped a Century and a Company*. Upper Saddle River, NJ: IBM Press.

Marcus, Clare Cooper. 1997. *House as a Mirror of Self: Exploring the Deeper Meaning of Home*. Reed Wheel and Weiner.

Marcus, George E. 1999. Critical Anthropology Now: An Introduction. In G. E. Marcus, ed., *Critical Anthropology Now: Unexpected Contexts, Shifting Constituencies, Changing Agendas*, pp. 3–28. Santa Fe, NM: School of American Research Press.

———. 1995. Ethnography in/of the World System: The Emergence of Multisited Ethnography. *Annual Review of Anthropology* 24:95–117.

Markowitz, G. and D. Rosner. 2002. *Deceit and Denial: The Deadly Politics of Industry Pollution*. Berkeley: University of California Press.

Martin, Emily. 1994. *Flexible Bodies: Tracking Immunity in American Culture from the Days of Polio to the Age of AIDS*. Boston: Beacon Press.

Marx, Karl. 1930 [1867] *Capital, vol 1*. I. E. and C. Paul, trans. 4th German ed. of Das Kapital (1890). London: Dent.

Mathews, Andrew S. 2011. *Instituting Nature: Authority, Expertise, and Power in Mexican Forests*. Cambridge, MA: MIT Press.

May, Mike. 1996. Risk Assessment: Bridging the Gap between Prediction and Experimentation. *Environmental Health Perspectives* 104(11):1150–51.

Mazurek, Jan. 1999. *Making Microchips: Policy, Globalization, and Economic Restructuring in the Semiconductor Industry*. Cambridge, MA: MIT Press.

McAfee, Noëlle. 2000. *Habermas, Kristeva, and Citizenship*. Ithaca: Cornell University Press.

McCarthy, Elise. 2012. Ethics and Ecologies: Negotiating Responsible and Sustainable Business in Ireland. PhD diss., Department of Anthropology, Rice University.

McCormick, S., P. Brown and S. Zavestoski. 2003. The Personal Is Scientific, The Scientific Is Political: The Public Paradigm of the Environmental Breast Cancer Movement. *Sociological Forum* 18:545–76.

McDaniels, Timothy and Mitchell J. Small, eds. 2004. *Risk Analysis and Society: An Interdisciplinary Characterization of the Field*. Cambridge, UK: Cambridge University Press.

McGraw, Thomas K., ed. 1997. *Creating Modern Capitalism*. Cambridge, MA: Harvard University Press.

McGurty, E. M. 2000. Warren County, NC, and the Emergence of the Environmental Justice Movement: Unlikely Coalitions and Shared Meanings in Local Collective Action. *Society and Natural Resources* 13:373–87.

Merleau-Ponty, Maurice. 2002 [1962]. *Phenomenology of Perception*. Forrest Williams, rev. and trans. New York: Humanities Press.

Meyer, John M. 2001. *Political Nature: Environmentalism and the Interpretation of Western Thought*. Cambridge, MA: MIT Press.

Michaels, D. 2008. *Doubt is Their Product: How Industry's Assault on Science Threatens Your Health*. New York: Oxford University Press.

Mitman, Gregg, Michelle Murphy, and Christopher Sellers. 2004. Introduction: A Cloud over History. In Gregg Mitman, Michelle Murphy, and Christopher Sellers, eds., *Landscapes of Exposure: Knowledge and Illness in Modern Environments*, pp.1–17. *OSIRIS* 19.

Moore, D. S. 1993. Contesting Terrain in Zimbabwe's Eastern Highlands: Political Ecology, Ethnography, and Peasant Resource Struggles. *Economic Geography* 69:380–401.

Morello-Frosch, R. 1997. Environmental Justice and California's "Riskscape": The Distribution of Air Toxics and Associated Cancer and Non-Cancer Risks among Diverse Communities. PhD diss., Department of Health Sciences, University of California, Berkeley.

Murphy, Michelle. 2006. *Sick Building Syndrome and the Problem of Uncertainty*. Durham, NC: Duke University Press.

———. 2004. Uncertain Exposures and the Privilege of Imperception: Activist Scientists and Race at the U.S. Environmental Protection Agency. *OSIRIS* 19:266–82.

Nader, Laura. 2011. Ethnography as Theory. *HAU: Journal of Ethnographic Theory* 1(1):211–19.

Nader, Laura, ed. 1996. *Naked Science: Anthropological Inquiry into Boundaries, Power, and Knowledge*. New York: Routledge.

Nash, June C. 1989. *From Tanker Town to High Tech: The Clash of Community and Industrial Cycles*. Albany: State University of New York Press.

National Academy of Public Administration. 2009. Putting Community First: A Promising Approach to Federal Collaboration for Environmental Improvement: An Evaluation of the Community Action for a Renewed Environment (CARE) Demonstration Program. NAPA Project Number 2100.

National Institute for Occupational Safety and Health. 2007. NIOSH Feasibility Assessment for a Cancer Study among Former IBM Employees Who Worked at the Endicott, New York Plant. March.

National Research Council. 1983. *Risk Assessment in the Federal Government: Managing the Process*. Washington, DC: National Academies Press.

Neumann, R. P. and R. A. Schroeder, eds. 1995. Manifest Ecological Destinies. *Antipode* 27(44):321–428.

New York State Department of Environmental Conservation Vapor Intrusion Guidance. Retrieved at http://www.dec.ny.gov/regulations/2588.html. Accessed October 13, 2008.

Northridge, Mary E. et al. 1999. Diesel Exhaust Exposure among Adolescents in Harlem: A Community-Driven Study. *American Journal of Public Health* 89(7):998–1002.

Novotny, P. 1995. Where We Live, Work and Play: Reframing the Cultural Landscape of Environmentalism in the Environmental Justice Movement. *New Political Science* 17(2):61–79.

O'Fallon, Liam and Allen Dearry. 2002. Community-Based Participatory Research as a Tool to Advance Environmental Health Sciences. *Environmental Health Perspectives* 110(2):155–59.

Olegario, R. 1997. IBM and the Two Thomas J. Watsons. In T. K. McGraw, ed., *Creating Modern Capitalism*, pp. 351–93. Cambridge, MA: Harvard University Press.

Oliver-Smith, Anthony. 1996. Anthropological Research on Hazards and Disasters. *Annual Review of Anthropology* 25:303–28.

Omohundro, Ellen. 2004. *Living in a Contaminated World: Community Structures, Environmental Risks and Decision Frameworks.* Burlington, VT: Ashgate.

Ong, Aihwa. 2006. *Neoliberalism as Exception.* Durham, NC: Duke University Press.

Ortner, Sherry B. 2006. *Anthropology and Social Theory: Culture, Power, and the Acting Subject.* Durham, NC: Duke University Press.

Ottinger, Gwen. 2013. *Refining Expertise: How Responsible Engineers Subvert Environmental Justice Challenges.* New York and London: New York University Press.

Ottinger, Gwen and Benjamin Cohen, eds. 2011. *Technoscience and Environmental Justice: Expert Cultures in a Grassroots Movement.* Cambridge, MA: MIT Press.

Peet, Richard, Paul Robbins, and Michael J. Watts, eds. 2011. *Global Political Ecology.* New York: Routledge.

Peirce, Charles S. 1934. How to Theorize. In *Collected Papers of Charles Sanders Pierce,* vol. 5, pp. 413–22). Cambridge, MA: Harvard University Press.

Pellow, David N. 2000. Environmental Inequality Formation: Toward a Theory of Environmental Injustice. *American Behavioral Scientist* 43(4):581–601.

Pellow, David N. and Robert J. Brulle, eds. 2005. *Power, Justice, and the Environment: A Critical Appraisal of the Environmental Justice Movement.* Cambridge, MA: MIT Press.

Pellow, David N. and Lee Sun-Hee Park. 2003. *The Silicon Valley of Dreams: Environmental Injustice, Immigrant Workers, and the High-Tech Global Economy.* New York: New York University Press.

Peluso, Nancy Lee. 1992. *Rich Forests, Poor People: Resource Control and Resistance in Java.* Berkeley: University of California Press.

Peluso, Nancy Lee and Michael Watts. 2001. Violent Environments. In N. L. Peluso and M. Watts, eds., *Violent Environments,* pp. 3–38. Ithaca: Cornell University Press.

Petryna, Adriana. 2002. *Life Exposed: Biological Citizenship after Chernobyl.* Princeton, NJ: Princeton University Press.

Pickering, Andrew, ed. 1992. *Science as Practice and Culture.* Chicago: University of Chicago Press.

Pirkey, Will M. 2012. Beyond Militant Particularisms: Collaboration and Hybridization in the "Contact Zones" of Environmentalism. *Capitalism Nature Socialism* 23(3):70–91.

Polish, David. 2009. Community Involvement Challenges at Vapor Intrusion Sites. Paper presented at the EPA National Forum on Vapor Intrusion. Philadelphia. January.

Porter, Theodore M. 1995. *Trust in Numbers: The Pursuit of Objectivity in Science and Public Life.* Princeton, NJ: Princeton University Press.

Prigogine, Ilya. 1997. *The End of Certainty: Time, Chaos, and the New Laws of Nature.* New York: Free Press.

Proctor, Robert. 1991. *Value-Free Science?* Cambridge, MA: Harvard University Press.

Pugh, Emerson W. 1995. *Building IBM: Shaping an Industry and Its Technology.* Cambridge, MA: MIT Press.

Pulido, L. 1996. *Environmentalism and Social Justice: Two Chicano Struggles in the Southwest.* Tucson: University of Arizona Press.

Rabinow, Paul. 2008. *Marking Time: On the Anthropology of the Contemporary.* Princeton, NJ: Princeton University Press.

———. 1999. *French DNA: Trouble in Purgatory.* Chicago: Chicago University Press.

Rajak, Dinah. 2011. *In Good Company: An Anatomy of Corporate Social Responsibility.* Stanford: Stanford University Press.

Rajan, Kaushik Sunder. 2006. *Biocapital: The Constitution of Postgenomic Life.* Durham, NC: Duke University Press.

Rajan, Kaushik Sunder, ed. 2012. *Lively Capital: Biotechnologies, Ethics, and Governance in Global Markets.* Durham, NC: Duke University Press.

Reagan, Ronald. 1984. Reagan-Bush Rally Speech. Union-Endicott High School. Endicott, New York. September 12.

Reed-Danahay, Deborah. 2005. *Locating Bourdieu.* Bloomington: Indiana University Press.

Reich, Robert. 2008. *Supercapitalism: The Transformation of Business, Democracy, and Everyday Life.* New York: Vintage.

Renfrew, Daniel. 2011. The Curse of Wealth: Political Ecologies of Latin American Neoliberalism. *Geography Compass* 5(8):581–94.

Renn, Ortwin. 2008. *Risk Governance: Coping with Uncertainty in a Complex World.* London: Earthscan.

Renn, Ortwin and Bernd Rohrmann. 2000. *Cross-Cultural Risk Perception: A Survey of Empirical Studies.* New York: Kluwer.

Reno, Joshua. 2011. Beyond Risk: Emplacement and the Production of Environmental Evidence. *American Ethnologist* 38(3): 516–30.

Residents Action Group of Endicott. RAGE Website. Retrieved at http://www.rage-ny.org. Accessed May 2, 2008.

Ricoeur, Paul. 2004. *Memory, History, Forgetting.* Kathleen Blamey and David Pellauer, trans. Chicago: University of Chicago Press.

Roberts, J. T. and M. M. Toffolon-Weiss. 2001. *Chronicles from the Environmental Justice Frontline*. Cambridge, MA: Cambridge University Press.

Royal Society. 1983. *Risk Assessment: Report of a Royal Society Study Group*. London.

Rustin, Michael. 1994. Incomplete Modernity: Ulrich Beck's Risk Society. *Radical Philosophy* 67. Summer Issue.

Rutherford, Danilyn. 2012. Kinky Empiricism. *Cultural Anthropology* 27(3):465–79.

Ryan, W. 1971. *Blaming the Victim*. New York: Pantheon.

Sandweiss, S. 1998. The Social Construction of Environmental Justice. In D. E. Camacho, ed., *Environmental Injustices, Political Struggles: Race, Class, and the Environment*, pp. 31–57. Durham, NC: Duke University Press.

Sanjek, Roger et al. 1990. *Fieldnotes: The Making of Anthropology*. Ithaca: Cornell University Press.

Scammell, Madeleine Kangsen et al. 2009. Tangible Evidence, Trust, and Power: Public Perceptions of Community Environmental Health Studies. *Social Science and Medicine* 68:143–53.

Schiller, Dan. 1999. *Digital Capitalism: Networking the Global Market System*. Cambridge, MA: MIT Press.

Schlosberg, David. 2007. *Defining Environmental Justice: Theories, Movements, and Nature*. New York: Oxford University Press.

———. 2004. Reconceiving Environmental Justice: Global Movements and Political Theories. *Environmental Politics* 13(3):517–40.

———. 1999. *Environmental Justice and the New Pluralism: The Challenge of Difference for Environmentalism*. New York: Oxford University Press.

Schumpeter, Joseph. 1942. *Capitalism, Socialism, and Democracy*. New York: Harper and Brothers.

Scott, C. S. and V. J. Cogliano. 2000. Introduction: Trichloroethylene Health Risks—State of the Science. *Environmental Health Perspectives* 108(S2):159–60.

Scott, James. 1998. *Seeing Like a State: How Certain Schemes to Improve the Human Condition Have Failed*. New Haven, CT: Yale University Press.

Sen, Amartya. 2009. *The Idea of Justice*. Cambridge, MA: Harvard University Press.

———. 2003. Foreword. In Paul Farmer, ed., *Pathologies of Power*, pp. xi–xvii. Berkeley: University of California Press.

Serres, Michel. 2011. *Malfeasance: Appropriation through Pollution?* Stanford: Stanford University Press.

———. 2011 [1974]. *Betrayal: The Thanatocracy*. In *Hermès III: La traduction*. Paris: Éditions de Minuit, pp. 73–104. Translated by Randolph Burks.

Shapin, Steven. 2008. *The Scientific Life: A Moral History of a Late Modern Vocation*. Chicago: University of Chicago Press.

Shepard, Peggy et al. 2002. Advancing Environmental Justice through Community Based Participatory Research. *Environmental Health Perspectives* 110(2):139–40.

Sherry, Susan et al. 1985. *High Tech and Toxics: A Guide for Local Communities*. Golden Empire Health Systems Agency.

Siegel, Lenny. 2012. *Vapor Intrusion Stakeholder-Involvement Forum.* Summary Notes. March.

———. 2010. Indoor air. TCE Blog posting. February 26.

———. 2009. A Stakeholder's Guide to Vapor Intrusion. Center for Public Environmental Oversight. November.

———. 2008. Report on the National Stakeholders' Panel on Vapor Intrusion. Center for Public Environmental Oversight. March.

———. 2007. Report on the Albany Vapor Intrusion Activists' Meeting. Center for Public Environmental Oversight. November.

Siegel, Lenny and John Markoff. 1985. *The High Cost of High Tech: The Dark Side of the Chip.* New York: Harper & Row.

Singer, Merrill. 2011. Down Cancer Alley: The Lived Experience of Health and Environmental Suffering in Louisiana's Chemical Corridor. *Medical Anthropology Quarterly* 25(2):141–63.

Slade, Giles. 2006. *Made to Break: Technology and Obsolescence in America.* Cambridge, MA: Harvard University Press.

Sloterdijk, Peter. 2011. *Bubbles: Spheres, Vol I: Microspherology.* Wieland Hoban, trans. Cambridge, MA: MIT Press.

Slovic, Paul. 2001. *The Perception of Risk.* London: Earthscan.

———. 1992. Perception of Risk: Reflections on the Psychometric Paradigm. In D. Golding and S. Krimsky, eds., *Theories of Risk*, pp. 117–52. London: Praeger.

———. 1987. Perception of Risk. *Science* 236(4799): 280–85.

Smith, Ted, David A. Sonnenfeld, and David N. Pellow, eds. 2006. *Challenging the Chip: Labor Rights and Environmental Justice in the Global Electronics Industry.* Philadelphia: Temple University Press.

Spears, Ellen. 2006. Toxic Knowledge: A Social History of Environmental Health in the New South's Anniston, Alabama, 1872–Present. PhD, diss., Department of American Studies, Emory University.

———. 2004. The Newtown Florist Club and the Quest for Environmental Justice in Gainesville, Georgia. In R. M. Packard, P. J. Brown, R. L. Berkelman, and H. Frumkin, eds., *Emerging Illnesses and Society: Negotiating the Public Health Agenda*, pp. 171–90. Baltimore, MD: Johns Hopkins University Press.

Stein, R., ed. 2004. *New Perspectives on Environmental Justice: Gender, Sexuality, and Activism.* New Brunswick, NJ: Rutgers University Press.

Stengers, Isabelle. 2011. *Cosmopolitics II.* Robert Bononno, trans. Minnesota: University of Minnesota Press.

———. 2010. *Cosmopolitics I.* Robert Bononno, trans. Minnesota: University of Minnesota Press.

Stephens, S. 2002. Bounding Uncertainty: The Post-Chernobyl Culture of Radiation Protection Experts. In. S. M. Hoffman and A. Oliver-Smith, eds., *Catastrophe and Culture: The Anthropology of Disaster.* Santa Fe, NM: School of American Research Press.

Strauss, Peter. 2010. Indoor air. TCE Blog posting. February 26.

Suarez-Villa, L. 2009. *Technocapitalism: A Critical Perspective on Technological Innovation and Corporatism*. Philadelphia: Temple University Press.

Sundberg, J. 2003. Strategies for Authenticity and Space in the Maya Biosphere Reserve, Petén, Guatemala. In K. S. Zimmerer and T .J. Bassett, eds., *Political Ecology: An Integrative Approach to Geography and Environment-Development Studies*, pp. 50–69. New York: Guilford Press.

Swedberg, Richard. 2012. Theorizing in Sociology and Social Science: Turning to the Context of Discovery. *Theory and Society* 41:1–40.

Szasz, A. 1994. *EcoPopulism: Toxic Waste and the Movement for Environmental Justice.* Minneapolis: University of Minnesota Press.

Tait, Joyce and Ann Bruce.2004. Global Change and Transboundary Risks. In T. McDaniels and M. J. Small, eds., *Risk Analysis and Society: An Interdisciplinary Characterization of the Field*, pp. 367–419. Cambridge, MA: Cambridge University Press.

Tarr, Joel A. 1996. *The Search for the Ultimate Sink: Urban Pollution in Historical Perspective*. Akron, OH: University of Akron Press.

Tarrow, Sidney. 1998 *Power in Movement: Social Movements and Contentious Politics.* Cambridge, UK: Cambridge University Press.

Taylor, D. 2000. The Rise of the Environmental Justice Paradigm: Injustice Framing and the Social Construction of Environmental Discourses. *American Behavioral Scientist* 43(4):508–80.

Tesh, Sylvia Noble.2000. *Uncertain Hazards: Environmental Activists and Scientific Proof.* Ithaca: Cornell University Press.

Tilt, Bryan. 2006. Perceptions of Risk from Industrial Pollution in China: A Comparison of Occupational Groups. *Human Organization* 65(2):115–27.

———. 2004. Risk, Pollution and Sustainability in Rural Sichuan, China. PhD diss., Department of Anthropology, University of Washington.

Traweek, Sharon. 1988. *Beamtimes and Lifetimes: The World of High Energy Physicists.* Cambridge, MA: Harvard University Press.

Tsing, Anna L. 2005. *Friction: An Ethnography of Global Connection.* Princeton, NJ: Princeton University Press.

Tuan, Yi-Fu. 1977. *Space and Place: The Perspective of Experience.* Minnesota: University of Minnesota Press.

———. 1974. *Topophilia: A Study of Environmental Perception, Attitudes, and Values.* New York: Columbia University Press.

Tucker, Pamela. 2000a. *Report of the Expert Panel Workshop on the Psychological Responses to Hazardous Substances.* Agency for Toxic Substances and Disease Registry, U.S. Department of Health and Human Services. Atlanta.

———. 2000b. Scientific Research Continues on the Psychological Responses to Toxic Contamination. *Hazardous Substances and Public Health* 10(1):1–11.

Turshen, Meredeth. 1977. The Political Ecology of Disease. *Review of Radical Political Economics* 9:45–60.

Uchitelle, Louis. 2004. *The Disposable American: Layoffs and Their Consequences*. New York: Knopf.

U.S. Environmental Protection Agency. 2012. *Conceptual Model Scenarios for the Vapor Intrusion Pathway*. Office of Solid Waste and Emergency Response. EPA 530-R-10-003. February.

———. 2008. *Indoor Air Vapor Intrusion Mitigation Approaches*. Office of Research and Development, National Risk Management Research Laboratory. EPA/600/R-08-115. October.

———. 2009. Proceedings: EPA National Forum on Vapor Intrusion. January 12–13. Philadelphia, PA.

———. 2007. Trichloroethylene (TCE): TEACH Chemical Summary. Retrieved at http://www.epa.gov/teach/chem_summ/TCE_summary.pdf. Accessed February 2, 2010.

———. 2004. *An Examination of EPA Risk Assessment Principles and Practices*. Prepared by the Risk Assessment Task Force. Washington, DC: Office of the Science Advisor.

———. 2001. *Trichloroethylene Health Risk Assessment: Synthesis and Characterization*. Office of Research and Development. EPA/600/P-01/002A. August.

———. 1994. *Model Standards and Techniques for Control of Radon in New Residential Buildings*. U.S. Environmental Protection Agency, Air and Radiation (6604-J). EPA 402-R-94-009. March.

———. 1993. *Air/Superfund National Technical Guidance Study Series: Options for Developing and Evaluating Mitigation Strategies for Indoor Air Impacts at CERCLA Sites*. EPA-451/R-93-012. September.

U.S. Environmental Protection Agency, Agency for Toxic Substances and Disease Registry. 1999. *A Citizen's Guide to Risk Assessments and Public Health Assessments at Contaminated Sites*. Washington, DC: U.S. EPA.

Visvanathan, Shiv. 2005. Knowledge, Justice and Democracy. In Melissa Leach, Ian Scoones, and Brian Wynne, eds., *Science and Citizens: Globalization and the Challenge of Engagement*, pp. 83–94. New York: Zed Books.

Wade, N. 2005. Geographic Society is Seeking a Genealogy of Humankind. *New York Times*, April 13: A16.

Wartenberg, D. D. Reyner, and C. S. Scott. 2000. Trichloroethylene and Cancer: Epidemiologic Evidence. *Environmental Health Perspectives* 108(supp2):161–76.

Weinberg, Alvin M. 1972. Science and Trans-science. *Minerva* 10:209–22.

West, Paige. 2005. Translation, Value, and Space: Theorizing an Ethnographic and Engaged Environmental Anthropology. *American Anthropologist* 107(4):632–42.

Whitehead, Alfred North. 1978. *Process and Reality*. New York: Macmillan.

———. 1933. *Adventures in Ideas*. New York: Free Press.

———. 1926. The Education of an Englishman. *Atlantic Monthly* 138:192.

Wilber, Tom. 2012. *Under the Surface: Fracking, Fortunes, and the Fate of the Marcellus Shale*. Ithaca: Cornell University Press.

Wilce, James M. 2008. Scientizing Bangladeshi Psychiatry: Parallelism, Enregisterment, and the Cure for a Magic Complex. *Language and Society* 37(1):91–114.

Wing, Steve et al. 1996. Community Based Collaboration for Environmental Justice: South-East Halifax Environmental Reawakening. *Environment and Urbanization* 8(2):129–40.

Wittgenstein, Ludwig. 1980 [1931]. *Culture and Value*. Peter Winch, trans. Chicago: University of Chicago Press.

Wolf, Eric R. 1999. *Envisioning Power: Ideologies of Dominance and Crisis*. Berkeley: University of California Press.

Wynne, Brian. 1989. Frameworks of Rationality in Risk Assessment. In J. Brown, ed., *Environmental Threats*, pp. 45–78. London: Frances Printer.

Yanchunas, Dom and Jeff Platsky. 2002. IBM-Endicott Site Sold, 2000 Jobs Appear Safe. *Press and Sun Bulletin*, July 1.

Zadek, Simon. 2001. *The Civil Corporation: The New Economy of Corporate Citizenship*. London: Earthscan.

Žižek, Slavoj. 2010. *Living in the End Times*. New York: Verso.

Zukin, Sharon. 1991. *Landscapes of Power: From Detroit to Disney World*. Berkeley: University of California Press.

Peter C. Little is Visiting Assistant Professor of Anthropology at the University of Louisville. He has published in a variety of academic journals, including *Medical Anthropology Quarterly, Journal of Political Ecology, Human Organization, Ethos,* and *Capitalism Nature Socialism.*

Printed in the United States
By Bookmasters